Pesticide Applications in Agricultural Systems

Pesticide Applications in Agricultural Systems

Editors

Giuseppe Manetto
Emanuele Cerruto
Domenico Longo
Rita Papa

MDPI • Basel • Beijing • Wuhan • Barcelona • Belgrade • Manchester • Tokyo • Cluj • Tianjin

Editors
Giuseppe Manetto
University of Catania
Italy

Emanuele Cerruto
University of Catania
Italy

Domenico Longo
University of Catania
Italy

Rita Papa
University of Catania
Italy

Editorial Office
MDPI
St. Alban-Anlage 66
4052 Basel, Switzerland

This is a reprint of articles from the Special Issue published online in the open access journal *Applied Sciences* (ISSN 2076-3417) (available at: https://www.mdpi.com/journal/applsci/special_issues/pesticide_agricultural).

For citation purposes, cite each article independently as indicated on the article page online and as indicated below:

LastName, A.A.; LastName, B.B.; LastName, C.C. Article Title. *Journal Name* **Year**, *Volume Number*, Page Range.

ISBN 978-3-0365-1303-4 (Hbk)
ISBN 978-3-0365-1304-1 (PDF)

Contents

About the Editors

Giuseppe Manetto, born on 2nd September 1965, graduated in Mechanical Engineering on 2nd April 1992, obtained his PhD in 1999, and is currently an Assistant Professor of Agricultural Mechanics at the Department of Agriculture, Food and Environment of the University of Catania. His continuous research activity, carried out with his participation in several research projects, covers various aspects of agricultural mechanics, such as the mechanical damage of fruit and vegetables in packing lines, safety aspects of workers in agricultural machines and agricultural and food plants, machinery for pesticide application, tillage and sod seeding, biogas plants, digestate spreading, and the development of prototypes (transportable dairy and milking plants for sheep and goats, mechanical distributor of arthropods, test bench for agricultural spray nozzles). He is the co-author of over 125 publications in national and international scientific journals and congress proceedings.

Emanuele Cerruto, born on 29 October 1964 in Modica (in the province of Ragusa, Italy), became a Researcher in 1994, obtained a PhD in 1995, and since 01/11/05, has been a full-time Associate Professor at the Di3A (Department of Agriculture, Food and Environment), section of Mechanics and Mechanization of the University of Catania (Italy). His research activity in the sector of machinery for pesticide application (foliar deposition, ground losses, worker exposure, spray quality), worker safety (noise and vibration exposure), post-harvest technology (packing lines, mechanical fruit damage) and renewable energies (anaerobic digestion plants, digestate spreading), goes back more than 25 years and is attested by more than 185 scientific papers and by the participation, as a member or responsible, of several units of research.

Domenico Longo, born on 14 February 1972 in Catania (Italy), obtained his degrees in Electronics and Automation Engineering in 2001 from University of Catania. In February 2005, he obtained his PhD in Electronics and Automation. From 2000, he has been involved in many robotics research projects at the Robotics Laboratory of University of Catania (Italy). His main research interests included mobile robots and, in particular, climbing robots for wall inspection. Since 31/12/2011 is a full-time Assistant Professor at the Di3A (Department of Agriculture, Food and Environment), Section of Mechanics and Mechanization of the University of Catania (Italy). His research activity is in the sector of robotics and mechatronics in agriculture, as well as the development of environmental sensors devices and environmental gas sensing apparatus. His activity is proven by many different scientific papers (https://orcid.org/0000-0001-9858-3061) and by the participation, as a member or responsible, in several units of research.

Rita Papa was born on 8 September 1982 in Sicily (Italy). In 2009, she graduated from University of Catania (Italy) in Agricultural Sciences and Technologies, specialising in engineering and economic aspects. From the same university, in 2013, she completed her PhD in Agricultural Engineering and, since then, she has been involved in research projects at the Di3A (Department of Agriculture, Food and Environment) of the University of Catania. Her research activity focused at first on different techniques for estimating evapotranspiration over Mediterranean crops; then, it dealt with spray quality in pesticide application, the mechanical distribution of beneficials, worker vibration exposure, anaerobic digestion plants and digestate spreading. She is a research fellow at the Di3A, Section of Mechanics and Mechanization, of the University of Catania.

 applied sciences

MDPI

Editorial

Special Issue on Pesticide Applications in Agricultural Systems

Giuseppe Manetto *, Emanuele Cerruto, Domenico Longo and Rita Papa

Department of Agriculture, Food and Environment (Di3A), Section of Mechanics and Mechanization, University of Catania, via Santa Sofia, 100, 95123 Catania, Italy; emanuele.cerruto@unict.it (E.C.); domenico.longo@unict.it (D.L.); rita.papa@gmail.com (R.P.)
* Correspondence: giuseppe.manetto@unict.it; Tel.: +39-095-7147-515

1. Introduction

The European Directive 2009/128/EC on the Sustainable Use of Pesticides recognizes the use of Plant Protection Products (PPP) as having great impact on human health and the environment [1]. The directive imposes several obligations, among which are the following: to inform the general public of the overall impacts of the use of pesticides and end-users on their safe handling and storage, as well as on the disposal of the empty packaging, tank mixtures, and remnants; to promote research programs aimed at determining the impacts of pesticide use; to pay particular attention to avoiding the pollution of surface and groundwater, the aquatic environment being especially sensitive to pesticides; to prohibit aerial spraying of pesticides due to its potential to cause significant adverse impacts, in particular from spray drift; to set up systems training for distributors, advisors, and professional users of pesticides; and to provide for systems for regular technical inspection of pesticide application equipment already in use.

Given this framework and, however, taking into account the benefits of pesticides when used rationally and carefully in increasing crop yields, improving food safety, quality of life and longevity, energy use, and environmental degradation [2], researchers, chemical companies, and sprayer manufacturers operate to make PPP application more sustainable, making more selective active ingredients available and promoting alternative approaches such as organic farming [3,4].

Citation: Manetto, G.; Cerruto, E.; Longo, D.; Papa, R. Special Issue on Pesticide Applications in Agricultural Systems. *Appl. Sci.* **2021**, *11*, 3695. https://doi.org/10.3390/app11083695

Received: 22 March 2021
Accepted: 16 April 2021
Published: 20 April 2021

2. Main Aspects Involved in Pesticide Application in Agriculture

Effective and efficient application of PPP in agriculture is a very complex task involving a lot of aspects ranging from active substance properties (physical status, volatility, selectivity, persistency) to climatic conditions (wind speed, air temperature, and relative humidity), from target structure (canopy or ground, vegetative development, canopy arrangement) to equipment features (state of maintenance, air assistance, proper calibration), and from operator awareness (level of education, training, expertise in choosing the appropriate operating parameters) to the reference legislative framework (ISO standards, European Directives, national and local obligations).

All these aspects, and many others, have been studied by the scientific community and are well referenced in the literature, and some of them have been treated by the authors that contributed to this Special Issue (SI) on Pesticide Applications in Agricultural Systems.

When spraying liquid PPP, a key factor that affects most aspects of a phytosanitary treatment is the drop size spectrum, because an optimal droplet spectrum ensures the transfer of the required dose to the target; minimizes the off-target losses due to evaporation, drift, and run-off; and reduces worker's dermal and inhalation exposure. The first four papers in this Special Issue focus on the importance of drop size spectrum on pesticide applications.

The first article, authored by Longo et al., presents a laboratory low-cost test bench for testing agricultural spray nozzles [5]. The paper provides a comprehensive description of the test bench, its design, construction, advantages, and limitations during its functioning, as well as the results of some trials with an ATR80 orange hollow cone nozzle to test the functionality of the whole system. The measurement method is inspired by the ISO 5682-1 [6] and is based on the image analysis of droplets trapped inside a layer of silicone oil of suitable density and viscosity.

The second article in this Special Issue, authored by Grella et al., is strictly connected to the first one. It deals with the designing of two pneumatic spouts for vineyard applications capable of generating droplet size spectra in a wide range of options [7]. Through laboratory trials simulating a real sprayer, the paper assesses the influence of hose insertion position, liquid flow rate, and airflow speed on droplet size, homogeneity, and driftability. A proper combination of these parameters allows for varying the spray quality from very fine to coarse/very coarse, providing farmers with a wide range of options to match the drift-reducing EU policy requirements when using pneumatic sprayers and the treatment specifications for every spray application.

The third article in this Special Issue, authored by Ranjan et al., exploits the use of modified micro-emitters, usually used in greenhouse irrigation systems, in fixed agrochemical delivery systems composed of a network of micro-sprayers/nozzles distributed in an apple orchard, replacing conventional hollow cone nozzles [8]. Ranjan et al. found that all the usually assessed spray parameters, i.e., spray coverage, canopy deposition, spray uniformity, and ground and aerial spray losses, were all affected by droplet spectrum and worked better with the modified irrigation micro-emitters when compared to the hollow cone nozzles. In addition, the installation cost was greatly reduced.

The fourth article in this Special Issue, authored by Wang et al., evaluates the handler's exposure to pesticides from stretcher-type power sprayers in orchards [9]. Spray drop spectrum also plays a key role in operator dermal and inhalation exposure, together with the task being performed, target features, canopy structure, the amount of pesticide handled, the duration of the activity, the personal protective equipment used, the type of equipment used and its status, and others [10]. Wang et al. found that inhalation exposure was negligible compared with dermal exposure, and that hands were the most exposed body part. Their results can be used as a reference for the handler's safety in pesticide management and mechanical orchard management.

The fifth paper in this Special Issue, authored by Rhoades and Stoddard, deals with the environmental effects of pesticides on non-target organisms [11]. Specifically, the research highlights a relationship between the spray application of a broad-spectrum insecticide and its effects on the predator activity of spiders and mosquito populations after treatment. Even if the pesticide dose was not lethal for spiders, their predator activity was reduced, and the mosquito population rebounded quickly after spray application. The authors propose further studies to assess the significance of the temporal association between the loss-of-predator ecosystem function of spiders and the rapid population rebound of mosquitos, as well as the adoption of other mosquito control methods, such as bacterial larvicides capable of avoiding these potential problems.

The last paper, authored by Facchinetti et al., is an example of how artificial intelligence can be used in precision spraying to reduce the volume sprayed and, subsequently, decrease the environmental impact of pesticides [12]. In the paper, the authors present the design of a prototype "rover" capable of adjusting the spraying parameters according to the target features, which are detected in real time by means of a 3D camera. Compared to the commonly distributed mixture volume in greenhouse fresh-cut salads, the prototype was able to reduce the amount of product sprayed by up to 55%, still assuring excellent crop coverage.

3. Future Perspectives on Pesticide Application in Agriculture

A continuous evolution is influencing spraying techniques to make pesticide application more sustainable; the monitoring of spray drift, using sensors to detect the target, the mapping of crop parameters, and developing variable rate sprayers are all aspects analyzed by researchers and sprayer manufacturers to implement sustainable precision farming systems. The availability of a huge amount of data coming from several sources (satellites, drones, ground) on all variables that can affect PPP distribution (climatic conditions, crop features, and the positioning and control of the sprayer) would result in the better targeted use of all available pest control measures. Therefore, as stated by the 2009/128/EC Directive, "it would contribute to a further reduction of the risks to human health and the environment and the dependency on the use of pesticides".

Funding: This research was carried out within the research project "Contributo della meccanica agraria e delle costruzioni rurali per il miglioramento della sostenibilità delle produzioni agricole, zoo-tecniche e agro-industriali. WP1: Impiego sostenibile di macchine irroratrici in serra e in pieno campo", financed inside the "Linea di intervento 2-dotazione ordinaria per attività istituzionali dei dipartimenti-2016-18-II annualità" of the University of Catania.

Acknowledgments: We are grateful to all authors for their contributions and making this Special Issue a success. Many thanks to all reviewers that provided responsible suggestions and comments. We would also like to express our gratitude to all members of the editorial team of Applied Sciences and especially to Christy Cui for their professional support.

Conflicts of Interest: The authors declare no conflict of interest.

References

1. European Union. Directive 2009/128/EC of the European Parliament and of the Council of 21 October 2009 establishing a framework for Community action to achieve the sustainable use of pesticides. *Off. J. L* **2009**, *309*, 71–86.
2. Cooper, J.; Dobson, H. The benefits of pesticides to mankind and the environment. *Crop. Prot.* **2007**, *26*, 1337–1348. [CrossRef]
3. Lamichhane, J.R. Pesticide use and risk reduction in European farming systems with IPM: An introduction to the special issue. *Crop. Prot.* **2017**, *97*, 1–6. [CrossRef]
4. Papa, R.; Manetto, G.; Cerruto, E.; Failla, S. Mechanical distribution of beneficial arthropods in greenhouse and open field: A review. *J. Agric. Eng.* **2018**, *785*, 81–91. [CrossRef]
5. Longo, D.; Manetto, G.; Papa, R.; Cerruto, E. Design and Construction of a Low-Cost Test Bench for Testing Agricultural Spray Nozzles. *Appl. Sci.* **2020**, *10*, 5221. [CrossRef]
6. ISO (International Organization for Standardization). *ISO 5682-1, Equipment for Crop. Protection—Spraying Equipment—Part 1: Test. Methods for Sprayer Nozzles*; ISO: Geneva, Switzerland, 2017. Available online: https://www.iso.org/standard/60053.html (accessed on 16 July 2020).
7. Grella, M.; Miranda-Fuentes, A.; Marucco, P.; Balsari, P.; Gioelli, F. Development of Drift-Reducing Spouts for Vineyard Pneumatic Sprayers: Measurement of Droplet Size Spectra Generated and Their Classification. *Appl. Sci.* **2020**, *10*, 7826. [CrossRef]
8. Ranjan, R.; Sinha, R.; Khot, L.; Hoheisel, G.; Grieshop, M.; Ledebuhr, M. Spatial Distribution of Spray from a Solid Set Canopy Delivery System in a High-Density Apple Orchard Retrofitted with Modified Emitters. *Appl. Sci.* **2021**, *11*, 709. [CrossRef]
9. Wang, Z.; Meng, Y.; Mei, X.; Ning, J.; Ma, X.; She, D. Assessment of Handler Exposure to Pesticides from Stretcher-Type Power Sprayers in Orchards. *Appl. Sci.* **2020**, *10*, 8684. [CrossRef]
10. Cerruto, E.; Manetto, G.; Santoro, F.; Pascuzzi, S. Operator dermal exposure to pesticides in tomato and strawberry greenhouses from hand-held sprayers. *Sustainability* **2018**, *10*, 2273. [CrossRef]
11. Rhoades, S.; Stoddard, P. Nonlethal Effects of Pesticides on Web-Building Spiders Might Account for Rapid Mosquito Population Rebound after Spray Application. *Appl. Sci.* **2021**, *11*, 1360. [CrossRef]
12. Facchinetti, D.; Santoro, S.; Galli, L.; Fontana, G.; Fedeli, L.; Parisi, S.; Bonacchi, L.; Šušnjar, S.; Salvai, F.; Coppola, G.; et al. Reduction of Pesticide Use in Fresh-Cut Salad Production through Artificial Intelligence. *Appl. Sci.* **2021**, *11*, 1992. [CrossRef]

Article

Design and Construction of a Low-Cost Test Bench for Testing Agricultural Spray Nozzles

Domenico Longo, Giuseppe Manetto *, Rita Papa and Emanuele Cerruto

Department of Agriculture, Food and Environment (Di3A), Section of Mechanics and Mechanization, University of Catania, via Santa Sofia, 100, 95123 Catania, Italy; domenico.longo@unict.it (D.L.); rita.papa@gmail.com (R.P.); emanuele.cerruto@unict.it (E.C.)

* Correspondence: giuseppe.manetto@unict.it; Tel.: +39-09-5714-7515

Received: 3 July 2020; Accepted: 27 July 2020; Published: 29 July 2020

Featured Application: A low-cost test bench to evaluate spray quality of agricultural nozzles was designed and built. The system allows for testing nozzles under effective work conditions at different pressure and flow rate values.

Abstract: Droplet size distribution is probably the most important feature of a spray as it affects all aspects of a phytosanitary treatment, i.e., biological, environmental, and safety aspects. This study describes a low-cost laboratory test bench able to analyze agricultural spray nozzles under realistic conditions. The design of the equipment was mainly based on the ISO 5682-1 standard. It has a couple of 3 m long rails, along which the nozzle under test moves while spraying, controlled by a closed-loop position and speed controller. The drops were captured with three Petri dishes containing silicone oil, photographed by means of a digital single-lens reflex (DSLR) camera, and then analyzed with the ImageJ software in order to measure the usual spray parameters: the volumetric diameters, the Sauter mean diameter, and the number mean diameter. Spray trials and tuning of the system parameters were managed by means of a purposely designed user interface running on a Windows 10 PC. Some tests were carried out by using an Albuz ATR80 orange hollow cone nozzle at the working pressures of 0.3, 0.5, 1.0, and 1.5 MPa. The results about spray quality agree with the factory information, and the whole system, even if some aspects still need improvements, has proven reliable.

Keywords: nozzle testing apparatus; pesticide; drop pulverization; drop size distribution; image analysis

---

1. Introduction

Spraying plant protection products (PPPs) is recognized as one of the agricultural activities most impacting on human health and the environment, also at the regulatory level [1]. Effects of PPPs on human health involve operators, farm workers and bystanders, consumers, professionals, and the general population, whereas those on the environment involve water, soil and atmosphere quality, and nontarget organisms such as vertebrates, useful arthropods, and other invertebrates. Therefore, on one hand researchers and manufacturers operate to get PPP application more sustainable according to the precision agriculture principles and on the other to promote nonchemical approaches.

Precision agriculture requires treating huge amount of data coming from several sources and sensors [2–7], from which to obtain prescription maps to be used with innovative spraying systems [8–16]. In addition, approaches for a more reasonable and sustainable use of PPPs include integrated pest management and organic farm [1,17–23], which allow for a reduction in the amount of chemical pesticides used.

Worker's exposure and environmental effects of pesticides are affected by various variables, among which include active substances, adjuvants, formulation, type of equipment used and its status, task being performed, target features, canopy structure, amount of pesticide handled, packaging, environmental conditions, duration of activity, personal protective equipment used, and others. A key factor is the spray spectrum, evaluated in terms of droplet size distribution [24–29], because an optimal droplet spectrum ensures the transfer of the required dose to the target, minimizes the off-target losses due to evaporation, drift, and run-off, and reduces dermal and inhalation worker exposure [30–37].

According to Schick [38], drops can be sampled for measurement purposes via spatial or flux techniques, and both methods affect the results. Spatial technique presupposes sampling instantaneously a number of drops occupying a given volume, whereas flux technique presupposes the examination of individual drops passing during an interval of time through the cross-section of a sampling region. The first method is sensitive to the number of particles in each class size and per unit volume, whereas the second to the particle flux.

Drop size analyzers available on the market that implement the spatial sampling technique include the optical imaging (OI) and the laser diffraction (LD) analyzers. Both systems are nonintrusive and do not influence the spray behavior during the measurement. The flux sampling technique is implemented by optical array probes (OAPs) and phase doppler particle analyzers (PDPAs). Both systems are nonintrusive also, are single particle counters, and allow for measuring drop size and velocity contemporarily.

Other methods, both intrusive and nonintrusive, are based on digital image analysis (DIA). Nonintrusive methods include the acquisition and analysis of the spray jet image [39], high-speed imaging [40], and shadowgraphy [41]. Among intrusive methods based on DIA, the use of water sensitive papers (WSPs) is probably the most diffused [42–49]. WSPs are artificial targets with a yellow surface layer that turns blue when in contact with water. This property allows for registering the droplet stains in spray tests. Assuming as known the spread factor (ratio between stain diameter and drop diameter) and negligible the overlap between stains, stain image analysis allows for determining the droplet size distribution.

A different approach, intrusive too, is to trap the droplets inside a layer of silicone oil of suitable density and viscosity and then analyze their image. The method is adopted by ISO 5682-1 [50], and it is exploited in this study to design and construct a low-cost test bench useful for evaluating agricultural spray nozzles.

The main aim of the paper is to provide a comprehensive description of the test bench (a preliminary study was presented in [51]), namely its design, construction, advantages, and limitations during its functioning. In addition, to test the functionality of the whole system, the results of some trials with an ATR80 orange hollow cone nozzle are reported.

2. Materials and Methods

2.1. Test Bench Design Guidelines

The test bench, applied on a movable trolley, was designed and built at the Section of Mechanics and Mechanization of the Di3A (University of Catania, Catania, Italy) based on the ISO 5682-1 standard, dealing with Equipment for crop protection–Spraying equipment–Test methods for sprayer nozzles. In addition, two main aspects were taken into consideration: the low cost of the whole equipment and test conditions similar to those present in standard commercial sprayers. With respect to the first aspect, the cost of mechanical structure and hardware components was less than 5000 €, whereas the cost of the image acquisition system (a camera with a macro lens that can also be used independently for other applications) was about 1200 €. Globally, the cost of the whole equipment was much lower than that of PDPA systems, though these systems allow for measuring drop velocity also. With regard to the second aspect, the test bench allows for testing agricultural spray nozzles under ordinary work conditions in a 1:1 scale for water flux, pressure, nozzle speed, and distance from the target.

According to the ISO 5682-1 reference, the nozzle under test should move above a row of Petri dishes with equal surface areas, spraying a test liquid (clean water with the addition of a soluble coloring agent, free from solids in suspension). Each Petri dish receives some of the droplets from the jet. All the droplets in each Petri dish should be measured and classified by size so that the cumulative volumetric curves and all the usual distribution parameters can be calculated. The accuracy of the equipment in measuring the drop diameter should be within 10 μm.

The selected nozzle speed should allow for a sufficient number of droplets to be collected (at least 2000 droplets should be collected to make a representative sample), while avoiding the merging of the droplets. The maximum speed fixed by the ISO 5682-1 standard is 3 m/s. The distance between the nozzle and Petri dishes should correspond to the normal distance between the nozzles and the crop. Finally, the test has to be carried out by letting the spraying nozzle pass once over the row of Petri dishes.

Other solutions, such as those described by Ali et al. [52], were discarded because they are based on the use of water sensitive papers. Calculation of spray spectrum parameters by using water sensitive papers (WSPs) requires knowledge of the spread factor as well as the overlap between stains to be made negligible. The spread factor is reported in WSP data sheets, but it is calculated at well-defined conditions (water temperature 20 °C, relative humidity 40%, droplets reaching the WSP at sedimentary velocity), which are different from real conditions and thus greater errors are expected.

2.2. Hydraulic Component Design

According to the aims of the research, the design of the hydraulic circuit was carried out in a way that is very similar to the standard spraying system installed on commercial sprayers. For this reason, a 70-liter plastic tank and a diaphragm pump (AR 30, Annovi Reverberi, Reggio Emilia, Italy, able to ensure pressure values up to at least 3 MPa) were chosen. The pump is driven by a single phase 230 V AC, 2.2 kW induction motor with a gearbox. As the pump capacity is much higher than the flow rate of the nozzle under test, the surplus was recirculated to the main tank by means of a pressure regulator as it is in the sprayers to ensure continuous mixing of the liquid.

In addition, to take into account the different working pressure ranges of the nozzles under test, the circuit was designed with two pressure lines (Figure 1): a low (up to 0.6 MPa) and a high pressure line (up to 3 MPa). According to the needs, the required working pressure was established by manually selecting the proper line by acting on the appropriate ball valves and then by acting on the corresponding pressure regulator.

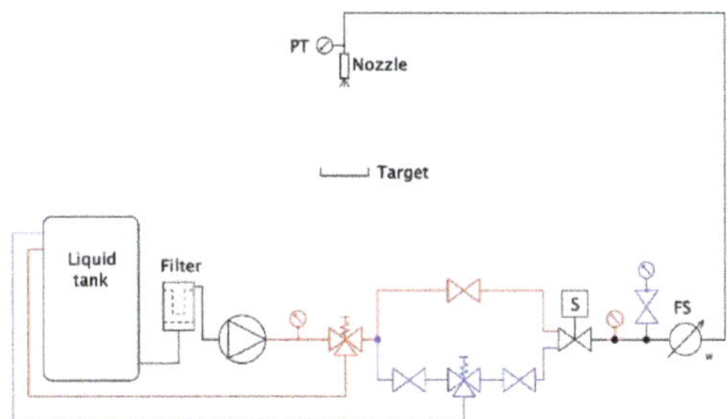

Figure 1. Scheme of the hydraulic circuit (blue: low pressure line; red: high pressure line; S: electro valve; FS: flux sensor; PT: pressure transmitter).

Due to the effect of the spray pressure and nozzle flow rate on the drop pulverization, the design of the hydraulic circuit also provided the sensors to measure in real time both quantities. Fluid pressure at the nozzle was measured by means of a piezoresistive pressure transmitter (Series 22 S model from Keller Italy Srl, Milano, Italy) and flow rate by means of a turbine flow sensor (SF800-6 model from Swissflow BV-Fazantlaan 4 6026 SN Maarheeze, The Netherlands). The pressure transmitter (3.0 MPa full-scale) has a standard 4–20 mA analog output, while the flow sensor has a digital pulse output with a nominal coefficient of 6000 pulse/L, a flow rate limit between 0.5 and 20 L/min, and a maximum operating pressure of 25 MPa.

2.3. Mechanical Component Design

The design of the mechanical components of the test bench took into consideration the following main aspects:

- possibility to control speed and position of the nozzle under test while spraying;
- mechanical insulation of the Petri dishes containing the drops from the pump and motors to avoid the transmission of vibration;
- general safety aspects due to the movement of the nozzle.

The movement of the nozzle under test was achieved by anchoring it to a mobile platform supported by four wheels and translating it along two rails mounted on the trolley by means of four steel supports (Figure 2). The mobile platform carries one triple nozzle holder, and the nozzle under test was manually selected. The platform was pulled by two parallel toothed belts (one at each side of the platform) of SynchroBelt^{TM} type, by means of a 250 W, 24 V DC permanent magnets (PMDC) brushed motor with a 6.75:1 gearbox. The rails, 3 m long, were spaced 0.6 m apart from each other, and were placed above and parallel to the trolley plane, at a distance of 0.6 m.

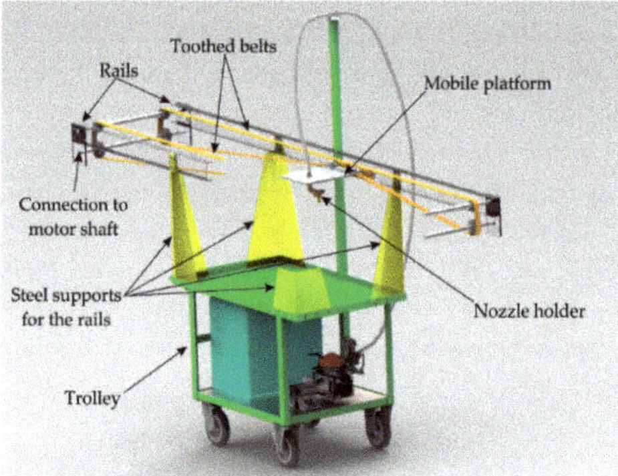

Figure 2. Partial section of the test bench showing its main components.

The motion of the mobile platform is closed-loop controlled by means of a motor control unit (MDC1460 model, Roboteq Inc., Scottsdale, AZ, USA). The acceleration, speed, and position are controlled by means of a 300 ppr (pulse per revolution) rotary quadrature optical encoder (ME22 model, INTECNO srl), directly applied to the main shaft of the PMDC motor. A simple symmetric trapezoidal speed shape was chosen, with the platform moving at constant speed while spraying above the Petri dishes. To obtain this speed profile vs. time, the initial and final acceleration/deceleration values were calculated with respect to the maximum desired speed in the flat part of the profile. The length of this central part was calculated to guarantee that the target will be sprayed at the desired speed. Calculations were carried out according to Equation (1):

$$\begin{cases} t_a = \frac{2L_a}{v_r} \\ t_r = \frac{L_r}{v_r} \\ a = \frac{v_r^2}{2L_a} \end{cases} \tag{1}$$

where t_a (s) is the acceleration/deceleration time (transient time), L_a (m) is the trails section covered by the nozzle at constant acceleration/deceleration, L_r (m) is the trails section covered by the nozzle at constant speed, v_r (m/s) is the desired constant nozzle speed, t_r is the time while the nozzle moves at constant speed (steady state time), and a (m/s^2) is the acceleration/deceleration value during transients.

The control unit implements, using a dedicated microcontroller, a speed and position profile control, by means of a proportional–integral–derivative (PID) control loop with an internal sampling rate of 1 kHz, using the encoder signal as feedback. The PID parameters were experimentally tuned by using the Ziegler–Nichols tuning method [53,54], in order to have a stable control system also at maximum speed and acceleration. The maximum speed of the platform, which can be kept constant while spraying over the target, is 1.5 m/s. Higher speeds would require acceleration and deceleration values not allowed for mechanical/electrical safety reasons; moreover, the length of the L_r section will shrink to very short and useless values. The motor control unit uses two limit switches at both ends of the rails in order to detect the correct home position and the end of the rails.

Due to the constraints imposed by the available space, only three Petri dishes were used to capture the drops, aligned parallel to the direction of travel of the nozzle while spraying. To avoid transmission of vibration that could affect drop size measurement, the Petri dishes were placed on a wood table, independent of the trolley and mechanically insulated from vibration sources (motors, pump). Drop image acquisition was carried out by means of a high-resolution camera. For this purpose, at the end of each experiment, the camera was applied to a suitable frame and manually hanged to the rails in fixed positions with respect to the Petri dishes.

2.4. Control Subsystem Design

A dedicated electrical cabinet was developed in order to accommodate all the necessary devices. In Figure 3, a block diagram is reported.

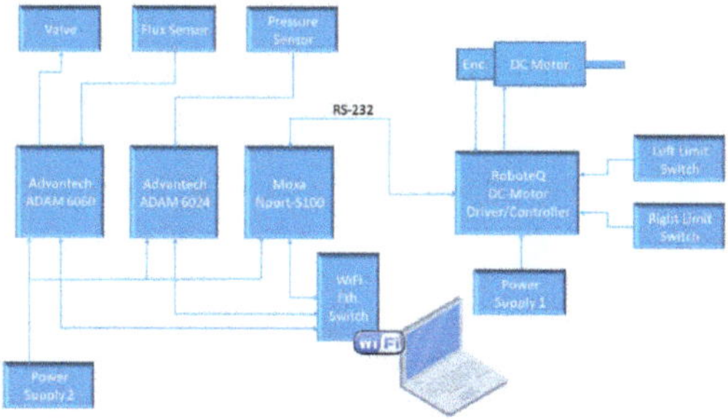

Figure 3. Block diagram of the control system architecture.

The PMDC motor for the mobile platform was controlled by a Roboteq MDC1460 controller. This controller is able to read any quadrature magnetic/optical encoder and handle two limit switches used to bound motor run between two ends. The controller is able to implement a PID closed-loop control over position/speed in an internal loop running at 1 kHz and is able to implement regenerative braking. This allows for better controller performance, but the power supply subsystem must be carefully designed in order to avoid over-voltage on the power supply rails. In battery-powered systems, the excess energy coming from the regenerative braking is used to recharge the batteries, but in systems powered from mains, this is not possible. To overcome this aspect, a low-cost, high-power, bidirectional TVS (transient voltage suppression) diode from Littelfuse (1.5KE30ca) was used to dissipate the braking back energy. The diode is able to safely dissipate 1500 W for about 1 ms and can handle about 200 A current flow for about 8 ms, with a response time of about 1 ps. These specifications are in excess to the estimated back energy coming from the programmed motor deceleration and also for the back energy coming from an emergency (immediate) motor stop. The motor controller is powered by a 480 W 230 V AC to 24 V DC AC/DC power supply unit (PSU). The power subsystem is completed by different fuses (with a high-power recirculating diode in parallel to the fuse to avoid system damages in case of a fuse blown) and a relay to allow the system to be selectively powered up (control subsystem and motor power subsystem). An emergency mushroom push-button with a mechanical key is provided for safety reasons. With this stop button, the system can be immediately powered down.

The control section, the sensors, and acquisition system, are powered up by a separate AC/DC PSU that is a 60 W 230 V AC to 12 V DC.

Figure 3 also describes the data acquisition and valve control subsystem. To acquire the two sensor signals (flow and pressure), two Advantech Ethernet modules with a maximum sample rate of 10 Hz were used: one ADAM6024 was used to read the output of the pressure transmitter (4–20 mA output) and one ADAM6060 to control a two-way electromagnetic valve, used to send the water flow to the nozzle. The valve has a 12 V DC 12 W coil and is actuated by means of a power relay. The same ADAM6060 module is able to read the output of the flow sensor that is typically a square wave signal with a frequency proportional to the flow rate. The ADAM6060 can read a maximum frequency of 3 kHz while the maximum expected frequency with a flow rate of about 1 L/min is about 100 Hz. The ADAM6060 can also be configured to count output pulses from the flowmeter, allowing us to measure the total water volume passed through the sensor. The ADAM modules have a standard 10/100 Base-TX Ethernet interface. They use different communication protocols based on Transmission Control Protocol/Internet Protocol (TCP/IP) and user datagram protocol (UDP). In this application, an ASCII UDP based protocol was selected.

The Roboteq controller has, among others, an RS232 interface able to send/receive configuration parameters as well as receive commands for the motor. In order to connect this serial interface with the other part of the system, an NPort 5100 (Moxa Inc., New Taipei City, Taiwan, R.O.C.) was used. This is a 1-port RS-232/422/485 serial device server able to act as a managed or transparent bridge between the RS232 bus and the Ethernet network. It has to be configured using its web interface and initialized using Telnet access; after that phase, it is possible to send and receive RS232 data/command to the Roboteq controller, using an UDP connection to the NPort 5100.

All the Ethernet devices are configured in a local network flat structure using the local IP address family 10.0.0.x/24. In order to allow the Windows 10 PC to have a direct connection to the system, a WiFi access point was used. In addition, the PC was configured on the same local network.

2.5. *Software User Interface Design*

The software user interface, running on a Windows 10 PC operating system, was designed to allow for managing the spray trials (i.e., starting and stopping spraying, nozzle speed, flow rate and pressure monitoring, positioning of the platform in a given location along the rails, stopping the experiment in case of safety problems) (Figure 4). It was built with Delphi Community Edition Integrated Development Environment (IDE) (Embarcadero Technologies Inc., Austin, TX, USA) using object-oriented Pascal language and the FireMonkey cross-platform graphic user interface (GUI) framework, developed also by Embarcadero Technologies Inc. All the commands were sent to the hardware layer through a user datagram protocol Ethernet packet at the different IP addresses assigned to each device, using a specific UDP port.

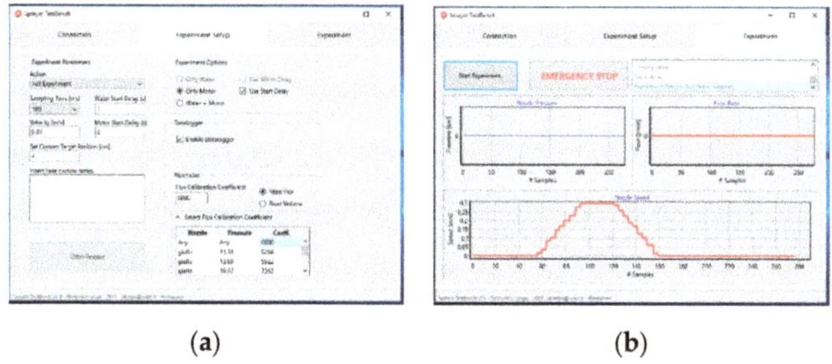

Figure 4. User interface: (**a**) tab used to manage the spray trials and set the working parameters; (**b**) tab used to monitor the sensor signals and to stop the system in case of emergency.

The application is composed by three main tabs. The "Connection" tab allows users to check the communication with each system device. Usually, just after the application start, a system check is performed automatically. In the tab there is a log box, in which all messages from/to the application and hardware devices are printed. In this way, it is possible to identify any system fault. In the "Experiment Setup" tab, it is possible to set up every aspect of the desired experiment, mainly the moving platform speed, some start delay for the platform motor and hydraulic electro valve, the system sampling time, to move the platform to the home position, and so on. Some options can be useful for testing purposes.

After setting up all the experiment parameters, using the "Experiment" tab it is possible to start the test; in this tab, it is possible to monitor different run-time graphs about the sensor data and the mobile platform speed profile. A log box in this tab is used to show the experiment progress until the end. The application also implements a data logger function that can save all the data and the experiment parameters in a text file using the standard comma separated value (CSV) format with time-stamp.

2.6. First Experimental Spraying Tests

These trials were carried out to test the functionality of the whole system and its sensitivity to pressure changes. No proper experimental design was defined, and results were considered as qualitative only. Future experiments and trials will be carried out and presented in other scientific papers, comparing several types of nozzles and testing under the same conditions the reference nozzles used to define the boundary regions of spray quality as recommended by the ISO/FDIS 25358:2018 standard [55]. Spraying tests were carried out with an ATR80 orange hollow cone nozzle (Albuz, France) at the working pressures of 0.3, 0.5, 1.0, and 1.5 MPa, and spraying as test liquid clean water with the addition of red Ponceau (Novema Srl, Torino, Italy) as a soluble coloring agent at the concentration of 2 g/L. In these first trials, surface tension of test liquid was not measured. Temperature and relative humidity of the ambient air were measured by using a thermohygrometer (HD 8901 model from Delta Ohm, Padova, Italy).

The three Petri dishes that were used to sample the drops had a diameter of 55 mm, and the distance between their centers was 195 mm (Figure 5). Nozzle speed during the tests was set to 1.5 m/s.

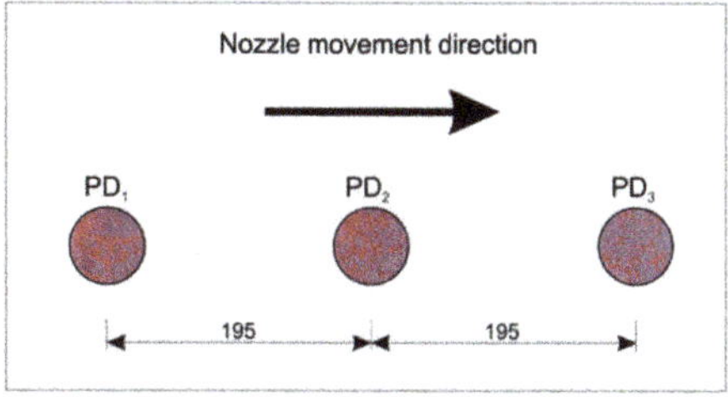

Figure 5. Positioning of the Petri dishes (PD) on the table of the test bench (sizes in mm).

Five milliliters of silicone oil (AR200 from Sigma-Aldrich, Milano, Italy) were deposited in each Petri dish by using a pipette (Mettler-Toledo, Milano, Italy), taking care to avoid bubble formation. Dynamic viscosity and volumetric mass of the silicone oil were 200 mPa s and 1050 kg/m^3, respectively.

After sprays, Petri dishes were photographed in situ by using a Nikon D5500 DSLR (digital single-lens reflex) camera equipped with a macro lens (Nikon Micro Nikkor AF-S 60 mm f/2.8 G ED) and an electronic flash (Neewer 48 Macro LED Ring Flash). To avoid involuntary movements, the camera was remotely controlled via the qDslrDashboard Version 3.5.3 application (http://dslrdashboard.info) running on an Android tablet. Spraying tests were repeated three times, so nine photos were taken for each working pressure. The images were saved as high-quality JPEG files with a resolution of 6000 × 4000 pixels. The red color of the test liquid allowed for an easy recognition of the droplets during the subsequent image analysis (Figure 6).

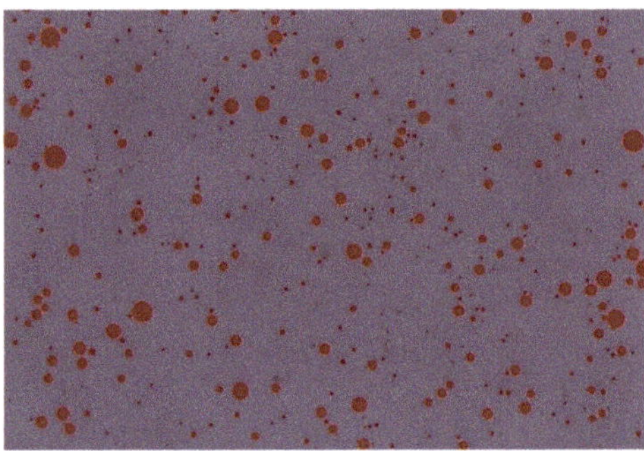

Figure 6. Droplets trapped into the silicone oil.

Images were calibrated by taking photos of a 10 × 10 mm grid pattern engraved on a glass disc, placed in correspondence of the three Petri dish positions, as far from the camera focal plane as the drops. The resulting scale factor ranged from 188.0 to 189.0 pixel/mm for the three Petri dish positions.

All images were analyzed with the ImageJ software, release 1.52a [56]. The use of this software for the spray droplet image analysis is well documented in the literature [44,45]. After conversion in 8-bit grey scale, images were segmented. The selected threshold value was obtained, on the basis of a careful inspection of the images, by increasing the "default" threshold value prompted by ImageJ by 20% on average. A threshold value directly linked to the grey level of the image as described in Sánchez-Hermosilla and Medina [57] was not exploited in this case because it was specifically studied to analyze water sensitive paper images, which are different from actual images, mainly due to the lighting conditions. In future trials, when other types of nozzles will be tested, a specific study will be dedicated to the segmentation process, and it will be checked whether the criterion adopted in this study for the ATR80 hollow cone nozzle is still valid. The ultimate goal will be to attempt to implement a segmentation procedure based on the correlation between some features of the images and the optimal threshold value, similar to that discussed in [57], in order to make the segmentation process as less subjective as possible.

After segmentation, images were processed with the watershed binary filter to separate some touching drops. The area (pixel) of the drops detected by ImageJ in each image was exported in text files for the subsequent analyses. Assuming that the measured area was circular, the droplet diameter was computed according to Equation (2):

$$D_i = \sqrt{\frac{4A_i}{\pi}} \tag{2}$$

where A_i (pixel) is the area of drop i. Particles with an area lower than 5 pixels were ignored. Droplet diameters expressed in pixels provided by Equation (2) were converted into real values by applying the proper scale factor, and then all the drop size distribution parameters were computed. The following quantities were calculated:

- D_{10}, as the arithmetic mean diameter;
- D_{20}, as the surface mean diameter;
- D_{30}, as the volume mean diameter;
- D_{32}, as the Sauter mean diameter (SMD), i.e., diameter of a drop having the same volume to surface area ratio as the total volume of all the drops to the total surface area of all the drops;
- $D_{v0.1}$, $D_{v0.5}$, and $D_{v0.9}$, as volumetric diameters, below which smaller droplets constitute, respectively, 10%, 50%, and 90% of the total volume;
- relative span factor (RSF), a dimensionless parameter indicative of the uniformity of the drop size distribution, defined as:

$$\mathrm{RSF} = \frac{D_{v0.9} - D_{v0.1}}{D_{v0.5}} \quad (3)$$

- number mean diameter (NMD), which is the droplet diameter below which the droplet diameter for 50% of the number of drops are smaller;
- V_{100} and V_{200}, as percentages of total volume of droplets smaller than, respectively, 100 and 200 μm in diameter.

All computations were carried out by means of custom functions written in R [58].

3. Results and Discussion

3.1. The Whole System Test

Figure 7 shows the main components of the entire system: the water tank with the electric motor pump and the manual pressure regulators; the mobile platform with the nozzle holder and the sensor pressure moving along the two rails; the cabinet with the electric/electronic components; the DSLR camera for drop image acquisition; and the wood table supporting the Petri dishes, which is mechanically insulated from the other parts of the frame.

At the beginning of the experiment, the electric motor driving the pump is switched on, and all the flow is recirculated to the tank (the electro valve of the hydraulic circuit is closed, Figure 1). After the working pressure is chosen and the manual pressure regulator is adjusted accordingly, the whole experiment is managed by the software. The user sets the desired platform speed (acceleration and deceleration are internally computed in the motor drive, according to Equation (1)), the initial delay necessary to reach steady-stay conditions, and the CSV file name to log the pressure, flow rate, velocity and position data. Then, the experiment starts: the mobile platform is carried at the home position (at one end of the rails), the electro valve is switched on, the nozzle starts spraying, and after the delay has elapsed, the platform moves and the test liquid is sprayed over the three Petri dishes containing silicone oil. After one pass, the platform stops at the opposite end of the rails. The distance between the spraying nozzle and Petri dishes is 0.5 m. The images of the drops trapped into the oil are immediately acquired in situ by using the camera.

3.2. Mechanical Tests

Figures 8 and 9 show the capabilities of the system to control speed and position. In particular, Figure 8 reports the position profiles vs. time recorded during the tests, while Figure 9 shows the speed profiles vs. position, directly obtained from the optical encoder. The profiles are almost superimposable for all the tests: The presence of the small quantities of noise is mainly due to limitations of the Windows operating system to accurately measure time intervals less than 100 ms. No malfunctioning was observed during all the experimental activities. In addition, the graphs show the platform reaching the reference speed (1.5 m/s) before approaching the first Petri dish, keeping it constant until going over the last one. The whole trajectory was completed in about 2.3 s.

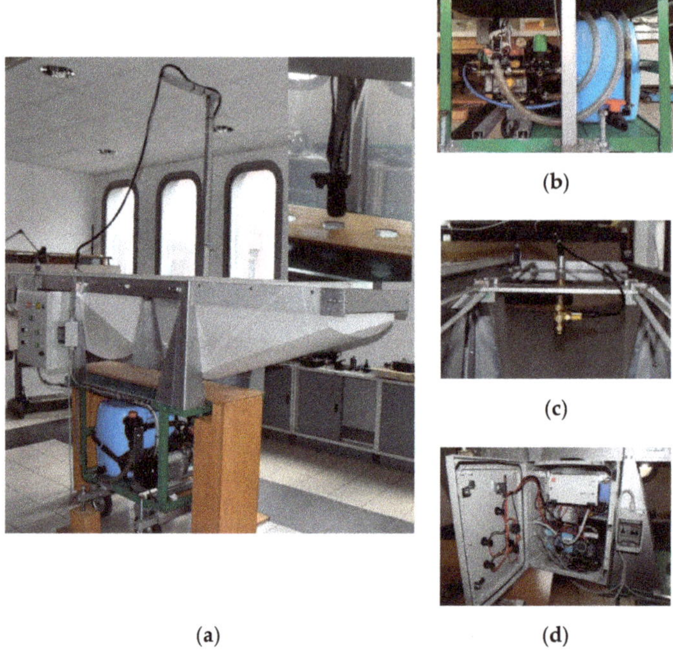

Figure 7. Main components of the test bench: (**a**) the whole system with the digital single-lens reflex (DSLR) camera and the wood table; (**b**) the main tank with the electric motor pump and the pressure regulators; (**c**) the mobile platform with the nozzle holder and the sensor pressure; (**d**) the cabinet with the electric components.

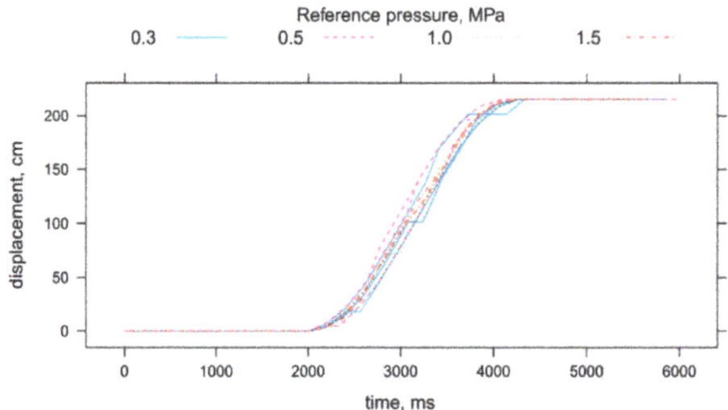

Figure 8. Position profiles of the mobile platform during the tests.

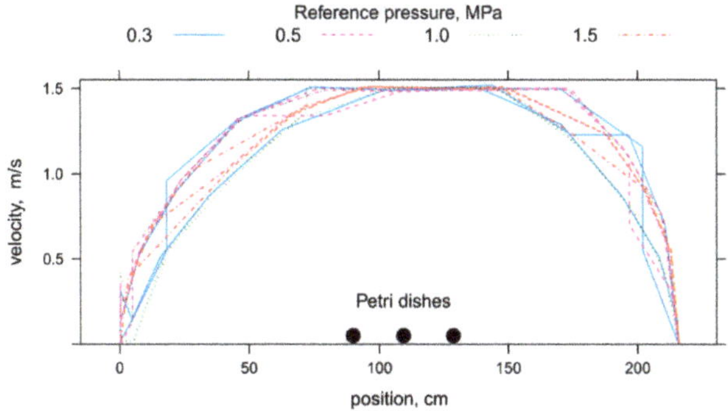

Figure 9. Speed profiles of the mobile platform during the tests.

3.3. Spraying Test Results

Air temperature and relative humidity during the experiments were 26 °C and 43%, respectively. The number of droplets detected in each of the analyzed images ranged from about 1700 to about 38,300. Taking into account all the images, the total number of sampled drops ranged from about 37,800 (0.3 MPa) to about 286,500 (1.5 MPa), much higher than the ISO 5682-1 requirements (representative samples composed by at least 2000 droplets).

Table 1 reports flow rate data for the nozzle under test, and Table 2 summarizes all the measured spray drop parameters in function of working pressures; mean values were computed assuming each Petri dish as an independent sample.

Table 1. Measured and reference flow rate values for the nozzle under test.

Measured Values		From Nozzle Data Sheets	
Pressure, MPa	**Flow Rate, L/min**	**Pressure, MPa**	**Flow Rate, L/min**
0.308	0.88		
0.517	0.98	0.5	0.99
1.036	1.37	1.0	1.39
1.492	1.63	1.5	1.69

Table 2. Spray drop parameters in function of working pressures.

	0.3 MPa		0.5 MPa		1.0 MPa		1.5 MPa	
Parameter	**Mean**	**Std Dev**	**Mean**	**Std Dev**	**Mean**	**Std Dev**	**Mean**	**Std Dev**
D_{10} (μm)	81.7	7.1	69.7	2.4	56.6	4.9	54.6	4.4
D_{20} (μm)	98.0	6.7	83.1	2.6	69.1	4.7	66.6	4.5
D_{30} (μm)	113.1	6.4	95.8	3.5	82.0	4.1	79.0	4.6
D_{32} (μm)	150.6	7.4	127.5	7.6	115.4	3.2	111.1	4.7
$D_{v0.1}$ (μm)	97.1	10.9	79.6	4.0	69.9	3.4	66.2	4.1
$D_{v0.5}$ (μm)	173.4	7.3	148.5	9.9	138.4	5.1	135.6	3.9
$D_{v0.9}$ (μm)	264.4	11.0	236.5	9.0	227.3	4.4	219.0	6.2
RSF	0.97	0.08	1.06	0.06	1.14	0.04	1.13	0.03
NMD (μm)	68.3	10.2	61.4	5.2	45.3	6.1	44.2	5.7
V_{100} (%)	11.6	3.3	22.0	4.4	25.8	2.6	28.7	2.9
V_{200} (%)	65.8	3.3	77.7	4.0	81.3	1.4	84.0	1.8

As expected, all diameters were affected by pressure: An increase in pressure determined a decrease in diameters, meaning a higher degree of drop pulverization. The measurement system was then able to correctly detect the influence of pressure changes. According to the Albuz catalogue, the spray quality of the ATR80 orange hollow cone nozzles at 0.5 MPa and over is classified as "very fine", based upon volumetric median diameter (VMD) $D_{v0.5}$ values lower than 159 μm, measured with phase Doppler anemometry (PDA) systems. Therefore, results obtained with the present test bench qualitatively agree with Albuz's information.

In addition, this measuring system was quite reliable and accurate: In fact, the coefficients of variation (CV) of all diameters (except $D_{v0.1}$ at 0.3 MPa and NMD) ranged between 1.9% and 8.6%. In the excluded cases, CVs ranged between 8.5% and 15.0%. As a general result, variability was higher when diameters were lower.

4. Conclusions

The design, development, and construction of a low-cost laboratory test bench suitable to analyze agricultural spray nozzles according to the ISO 5682-1 procedure were discussed in the paper. The preliminary tests show its usefulness under several aspects:

- The software interface to control the test bench and to set the test conditions is simple, and its organization in tabs (the first to set the test parameters; the second one to see graphically the evolution of the test variables) makes it easy to use the test bench.
- The hydraulic circuit allows for testing nozzles under work conditions similar to those present in commercial sprayers, using a standard diaphragm pump and standard manual pressure regulators.
- The electromechanic components allow for a fine and precise control of speed and position of the nozzle under test.
- The image acquisition system, based on a digital single-lens reflex camera, may be easily updated if higher resolutions are required. The actual system, with a scale factor of 188.0–189.0 pixel/mm, allows for detecting the drops whose diameter is greater than 10 μm, which can be considered suitable for similar applications.
- In measuring the single drop diameters, it is possible to devise the size diameter probability distribution function and all the usual spray drop parameters. The preliminary tests with the Albuz ATR80 orange hollow cone nozzles produced results in accordance with factory data sheets.
- Further tests that are aimed at comparing several nozzle types (hollow cone, fan, air induction) with the reference nozzles recommended by ISO/FDIS 25358:2018 [55] to define the boundaries/borders between the size classes may better assess the capabilities of the test bench.

The system also has some limitations due to the specific hardware implementation. The most severe limitation is imposed by the rail length, which does not allow for the maximum speed of the mobile platform, while spraying above the Petri dishes, to be higher than 1.5 m/s. Higher speed values to be kept constant while the nozzle sprays above the Petri dishes would require higher acceleration values, but the maximum acceleration is limited for mechanical/electrical safety reasons. Increasing the platform speed is a practical way to reduce overlaps between drops and then to simplify the subsequent image analysis procedure.

A minor limitation is related to the maximum sample rate of the Advantech Ethernet modules, used for monitoring the position and speed profile and for acquiring data from sensors. The trajectory control loop is achieved via hardware with a 1 kHz sampling frequency, but monitoring and logging operations are left to the Windows operating system, which cannot guarantee a constant sampling rate and good accuracy for time intervals less than 100 ms; thus, the speed and position graphs may result deformed. An optimization of the reading and writing routines and of data filtering may improve these aspects.

Finally, a significant improvement may be introduced in the camera positioning system. The camera is actually applied to a metallic frame, which in turn is manually hanged to the rails in predetermined positions. A better solution could be to hang the camera directly to the mobile platform after the spray passes and then to position it in correspondence of the Petri dishes by exploiting the control position system of the motor. This could speed up the image acquisition procedure and then reduce any effect of external factors.

Finally, taking into account all the above considerations, it is possible to conclude that the main objectives fixed were achieved with the test bench and that its modularity allows for further developments to overcome the limitations highlighted, without compromising the low cost.

Author Contributions: Conceptualization, E.C., D.L., G.M., and R.P.; methodology, E.C., D.L., G.M., and R.P.; software, D.L.; validation, E.C., D.L., G.M., and R.P.; formal analysis, E.C., D.L., G.M., and R.P.; investigation, E.C., D.L., G.M., and R.P.; data curation, E.C., D.L., G.M., and R.P.; writing—original draft preparation, E.C.; writing—review and editing, E.C., D.L., G.M., and R.P. All authors have read and agreed to the published version of the manuscript.

Funding: This research was carried out within the research project "Contributo della meccanica agraria e delle costruzioni rurali per il miglioramento della sostenibilità delle produzioni agricole, zootecniche e agro-industriali. WP1: Impiego sostenibile di macchine irroratrici in serra e in pieno campo", financed inside the "Linea di intervento 2-dotazione ordinaria per attività istituzionali dei dipartimenti-2016-18-II annualità" of the University of Catania.

Conflicts of Interest: The authors declare no conflict of interest.

Abbreviation

Symbol	
a	acceleration/deceleration value during transients, m/s^2
L_a	trails section covered by the nozzle at constant acceleration/deceleration, m
L_r	trails section covered by the nozzle at constant speed, m
t_a	acceleration/deceleration time (transient time), s
t_r	time while the nozzle moves at constant speed (steady state time), s
v_r	desired constant nozzle speed, m/s
A_i	area of droplet i detected by ImageJ, pixel
D_i	diameter of droplet i, pixel
D_{10}	arithmetic mean diameter, µm
D_{20}	surface mean diameter, µm
D_{30}	volume mean diameter, µm
D_{32}	Sauter mean diameter (SMD) or the diameter of a drop having the same volume to surface area ratio as the total volume of all the drops to the total surface area of all the drops, µm
$D_{v0.1}$, $D_{v0.9}$	volumetric diameters below which smaller droplets constitute, respectively, 10% and 90% of the total volume, µm
$D_{v0.5}$	volumetric median diameter (VMD), below which smaller droplets constitute 50% of the total volume, µm
NMD	number mean diameter, the droplet diameter below which the droplet diameter for 50% of the number of drops are smaller, µm
RSF	relative span factor, a dimensionless parameter indicative of the uniformity of the drop size distribution
SMD	Sauter mean diameter, µm
VMD	volume median diameter, µm
V_{100}, V_{200}	proportion of total volume of droplets smaller than, respectively, 100 and 200 µm in diameter, %

Acronym

AC	Alternated current
CSV	Comma separated value
CV	Coefficient of variation
DC	Direct current
DIA	Digital image analysis
DSLR	Digital single-lens reflex
GUI	Graphic user interface
IDE	Integrated development environment
IP	Internet protocol
LD	Laser diffraction
OAP	Optical array probes
OI	Optical imaging
PD	Petri dish
PDA	Phase Doppler anemometry
PDPA	Phase Doppler particle analyzers
PID	Proportional–integral–derivative
PMDC	Permanent magnets direct current
PPP	Plant protection products
PSU	Power supply unit
TCP	Transmission control protocol
TCP/IP	Transmission control protocol/Internet protocol

References

1. European Union. Directive 2009/128/EC of the European Parliament and of the Council of 21 October 2009 establishing a framework for Community action to achieve the sustainable use of pesticides. *Off. J. L* **2009**, *309*, 71–86.
2. Cox, S. Information technology: The global key to precision agriculture and sustainability. *Comput. Electron. Agric.* **2002**, *36*, 93–111. [CrossRef]
3. Zhang, N.; Wang, M.; Wang, N. Precision agriculture—A worldwide overview. *Comput. Electron. Agric.* **2002**, *36*, 113–132. [CrossRef]
4. Camilli, A.; Cugnasca, C.E.; Saraiva, A.M.; Hirakawa, A.R.; Corrêa, P.L.P. From wireless sensors to field mapping: Anatomy of an application for precision agriculture. *Comput. Electron. Agric.* **2007**, *58*, 25–36. [CrossRef]
5. Sankaran, S.; Mishra, A.; Ehsani, R.; Davis, C. A review of advanced techniques for detecting plant diseases. *Comput. Electron. Agric.* **2010**, *72*, 1–13. [CrossRef]
6. Gil, E.; Arnó, J.; Llorens, J.; Sanz, R.; Llop, J.; Rosell-Polo, J.R.; Gallart, M.; Escolà, A. Advanced technologies for the improvement of spray application techniques in Spanish viticulture: An overview. *Sensors* **2014**, *14*, 691–708. [CrossRef]
7. Jiao, L.; Dong, D.; Feng, H.; Zhao, X.; Chen, L. Monitoring spray drift in aerial spray application based on infrared thermal imaging technology. *Comput. Electron. Agric.* **2016**, *121*, 135–140. [CrossRef]
8. Escolà, A.; Rosell-Polo, J.R.; Planas, S.; Gil, E.; Pomar, J.; Camp, F.; Llorens, J.; Solanelles, F. Variable rate sprayer. Part 1—orchard prototype: Design, implementation and validation. *Comput. Electron. Agric.* **2013**, *95*, 122–135. [CrossRef]
9. Gil, E.; Llorens, J.; Llop, J.; Fàbregas, X.; Escolà, A.; Rosell-Polo, J.R. Variable rate sprayer. Part 2—Vineyard prototype: Design, implementation, and validation. *Comput. Electron. Agric.* **2013**, *95*, 136–150. [CrossRef]
10. Pascuzzi, S.; Cerruto, E. Spray deposition in "tendone" vineyards when using a pneumatic electrostatic sprayer. *Crop Prot.* **2015**, *68*, 1–11. [CrossRef]
11. Oberti, R.; Marchi, M.; Tirelli, P.; Calcante, A.; Iriti, M.; Tona, E.; Hočevar, M.; Baur, J.; Pfaff, J.; Schütz, C.; et al. Selective spraying of grapevines for disease control using a modular agricultural robot. *Biosyst. Eng.* **2016**, *146*, 203–215. [CrossRef]
12. Qin, W.C.; Qiu, B.J.; Xue, X.Y.; Chen, C.; Xu, Z.F.; Zhou, Q.Q. Droplet deposition and control effect of insecticides sprayed with an unmanned aerial vehicle against plant hoppers. *Crop Prot.* **2016**, *85*, 79–88. [CrossRef]

13. Grella, M.; Gallart, M.; Marucco, P.; Balsari, P.; Gil, E. Ground deposition and airborne spray drift assessment in vineyard and orchard: The influence of environmental variables and sprayer settings. *Sustainability* **2017**, *9*, 728. [CrossRef]
14. Pascuzzi, S.; Cerruto, E.; Manetto, G. Foliar spray deposition in a "tendone" vineyard as affected by airflow rate, volume rate and vegetative development. *Crop Prot.* **2017**, *91*, 34–48. [CrossRef]
15. Rao Mogili, U.M.; Deepak, B.B.V.L. Review on application of drone systems in precision agriculture. *Procedia Comput. Sci.* **2018**, *133*, 502–509. [CrossRef]
16. Pascuzzi, S.; Santoro, F.; Manetto, G.; Cerruto, E. Study of the correlation between foliar and patternator deposits in a "tendone" vineyard. *Agric. Eng. Int. CIGR J.* **2018**, *20*, 97–107.
17. Gan-Mor, S.; Matthews, G.A. Recent developments in sprayers for application of biopesticides—An overview. *Biosyst. Eng.* **2003**, *84*, 119–125. [CrossRef]
18. Blandini, G.; Emma, G.; Failla, S.; Manetto, G. A prototype for mechanical distribution of beneficials. *Acta Hortic.* **2008**, *2*, 1515–1522. [CrossRef]
19. Pezzi, F.; Martelli, R.; Lanzoni, A.; Maini, S. Effects of mechanical distribution on survival and reproduction of *Phytoseiulus persimilis* and *Amblyseius swirskii*. *Biosyst. Eng.* **2015**, *129*, 11–19. [CrossRef]
20. Lamichhane, J.R. Pesticide use and risk reduction in European farming systems with IPM: An introduction to the special issue. *Crop Prot.* **2017**, *97*. [CrossRef]
21. Papa, R.; Manetto, G.; Cerruto, E.; Failla, S. Mechanical distribution of beneficial arthropods in greenhouse and open field: A review. *J. Agric. Eng.* **2018**, *49*, 81–91. [CrossRef]
22. Lee, R.; den Uyl, R.; Runhaar, H. Assessment of policy instruments for pesticide use reduction in Europe; Learning from a systematic literature review. *Crop Prot.* **2019**, *126*, 104929. [CrossRef]
23. Mantzoukas, S.; Eliopoulos, P.A. Endophytic entomopathogenic fungi: A valuable biological control tool against plant pests. *Appl. Sci.* **2020**, *10*, 360. [CrossRef]
24. Hewitt, A.J. Droplet size and agricultural spraying, part I: Atomization, spray transport, deposition, drift, and droplet size measurement techniques. *At. Sprays* **1997**, *7*, 235–244. [CrossRef]
25. Matthews, G.A. How was the pesticide applied? *Crop Prot.* **2004**, *23*, 651–653. [CrossRef]
26. Nuyttens, D.; Baetens, K.; De Schampheleire, M.; Sonck, B. Effect of nozzle type, size and pressure on spray droplet characteristics. *Biosyst. Eng.* **2007**, *97*, 333–345. [CrossRef]
27. Garcerá, C.; Román, C.; Moltó, E.; Abad, R.; Insa, J.A.; Torrent, X.; Chueca, P. Comparison between standard and drift reducing nozzles for pesticide application in citrus: Part II. Effects on canopy spray distribution, control efficacy of *Aonidiella aurantii* (Maskell), beneficial parasitoids and pesticide residues on fruit. *Crop Prot.* **2017**, *94*, 83–96. [CrossRef]
28. Torrent, X.; Garcerá, C.; Moltó, E.; Chueca, P.; Abad, R.; Grafulla, C.; Planas, S. Comparison between standard and drift reducing nozzles for pesticide application in citrus: Part I. Effects on wind tunnel and field spray drift. *Crop Prot.* **2017**, *96*, 130–143. [CrossRef]
29. Torrent, X.; Gregorio, E.; Douzals, J.P.; Tinet, C.; Rosell-Polo, J.R.; Planas, S. Assessment of spray drift potential reduction for hollow-cone nozzles: Part 1. Classification using indirect methods. *Sci. Total Environ.* **2019**. [CrossRef]
30. Berger-Preiß, E.; Boehnckey, A.; Könnecker, G.; Mangelsdorf, I.; Holthenrich, D.; Koch, W. Inhalational and dermal exposures during spray application of biocides. *Int. J. Hyg. Environ. Health* **2005**, *208*, 357–372. [CrossRef]
31. Nuyttens, D.; Braekman, P.; Windey, S.; Sonck, B. Potential dermal pesticide exposure affected by greenhouse spray application technique. *Pest Manag. Sci.* **2009**, *65*, 781–790. [CrossRef] [PubMed]
32. Sánchez-Hermosilla, J.; Rincón, V.J.; Páez, F.; Agüera, F.; Carvajal, F. Field evaluation of a self-propelled sprayer and effects of the application rate on spray deposition and losses to the ground in greenhouse tomato crops. *Pest Manag. Sci.* **2011**, *67*, 942–947. [CrossRef] [PubMed]
33. Sánchez-Hermosilla, J.; Páez, F.; Rincón, V.J.; Carvajal, F. Evaluation of the effect of spray pressure in hand-held sprayers in a greenhouse tomato crop. *Crop Prot.* **2013**, *54*, 121–125. [CrossRef]
34. Gil, E.; Balsari, P.; Gallart, M.; Llorens, J.; Marucco, P.; Andersen, P.G.; Fàbregas, X.; Llop, J. Determination of drift potential of different flat fan nozzles on a boom sprayer using a test bench. *Crop Prot.* **2014**, *56*, 58–68. [CrossRef]
35. Cerruto, E.; Manetto, G.; Santoro, F.; Pascuzzi, S. Operator dermal exposure to pesticides in tomato and strawberry greenhouses from hand-held sprayers. *Sustainability* **2018**, *10*, 2273. [CrossRef]
36. Rincón, V.J.; Páez, F.C.; Sánchez-Hermosilla, J. Potential dermal exposure to operators applying pesticide on greenhouse crops using low-cost equipment. *Sci. Total Environ.* **2018**, *630*, 1181–1187. [CrossRef]

37. Lodwik, D.; Pietrzyk, J.; Malesa, W. Analysis of volume distribution and evaluation of the spraying spectrum in terms of spraying quality. *Appl. Sci.* **2020**, *10*, 2395. [CrossRef]
38. Schick, R.J. Spray Technology Reference Guide: Understanding drop size. In *Spraying Systems Bulletin no. 459C*; Spraying Systems Co.: Wheaton, IL, USA, 2008; Available online: https://www.spray.com/literature_pdfs/B459C_Understanding_Drop_Size.pdf (accessed on 11 November 2019).
39. Lad, N.; Aroussi, E.A.; Muhamad Said, M.F. Droplet size measurement for liquid spray using digital image analysis technique. *J. Appl. Sci.* **2011**, *11*, 1966–1972. [CrossRef]
40. De Cock, N.; Massinon, M.; Nuyttens, D.; Dekeyser, D.; Lebeau, F. Measurements of reference ISO nozzles by high-speed imaging. *Crop Prot.* **2016**, *89*, 105–115. [CrossRef]
41. De Cock, N.; Massinon, M.; Ouled Taleb Salah, S.; Mercatoris, B.C.N.; Lebeau, F. Droplet size distribution measurements of ISO nozzles by shadowgraphy method. *Commun. Appl. Biol. Sci.* **2015**, *80*, 295–301.
42. Hoffmann, W.C.; Hewitt, A.J. Comparison of three imaging systems for water-sensitive papers. *Appl. Eng. Agric.* **2005**, *21*, 961–964. [CrossRef]
43. Marçal, A.R.S.; Cunha, M. Image processing of artificial targets for automatic evaluation of spray quality. *Trans. ASABE* **2008**, *51*, 811–821. [CrossRef]
44. Zhu, H.; Salyani, M.; Fox, R.D. A portable scanning system for evaluation of spray deposit distribution. *Comput. Electron. Agric.* **2011**, *76*, 38–43. [CrossRef]
45. Cunha, M.; Carvalho, C.; Marcal, A.R.S. Assessing the ability of image processing software to analyse spray quality on water-sensitive papers used as artificial targets. *Biosyst. Eng.* **2012**, *111*, 11–23. [CrossRef]
46. Cerruto, E.; Aglieco, C.; Failla, S.; Manetto, G. Parameters influencing deposit estimation when using water sensitive papers. *J. Agric. Eng.* **2013**, *44*, 62–70. [CrossRef]
47. Salyani, M.; Zhu, H.; Sweeb, R.D.; Pai, N. Assessment of spray distribution with water-sensitive paper. *Agric. Eng. Int. CIGR J.* **2013**, *15*, 101–111.
48. Cerruto, E.; Failla, S.; Longo, D.; Manetto, G. Simulation of water sensitive papers for spray analysis. *Agric. Eng. Int. CIGR J.* **2016**, *18*, 22–29.
49. Cerruto, E.; Manetto, G.; Longo, D.; Failla, S.; Papa, R. A model to estimate the spray deposit by simulated water sensitive papers. *Crop Prot.* **2019**, *124*, 104861. [CrossRef]
50. ISO (International Organization for Standardization). *ISO 5682-1, Equipment for Crop Protection—Spraying Equipment—Part 1: Test Methods for Sprayer Nozzles*; ISO: Geneva, Switzerland, 2017. Available online: https://www.iso.org/standard/60053.html (accessed on 16 July 2020).
51. Cerruto, E.; Manetto, G.; Longo, D.; Failla, S.; Schillaci, G. A laboratory system for nozzle spray analysis. *Chem. Eng. Trans.* **2017**, *58*, 751–756. [CrossRef]
52. Ali, A.; Awan, A.; Khan, F.H.; Khan, M.A. Fabrication of Ultra Low Volume (ULV) pesticide sprayer test bench. *Pak. J. Agric. Sci.* **2011**, *48*, 135–140.
53. Ziegler, G.; Nichols, N.B. Optimum settings for automatic controllers. *Trans. ASME* **1942**, *64*, 759–768. [CrossRef]
54. Meshram, P.M.; Kanojiya, R.G. Tuning of PID controller using Ziegler-Nichols method for speed control of DC motor. In Proceedings of the IEEE-International Conference on Advances in Engineering, Science and Management (ICAESM-2012), Nagapattinam, Tamil Nadu, India, 30–31 March 2012; pp. 117–122.
55. ISO (International Organization for Standardization). *ISO/FDIS 25358:2018, Crop Protection Equipment—Droplet-Size Spectra from Atomizers—Measurement and Classification*; ISO: Geneva, Switzerland, 2018. Available online: https://www.iso.org/standard/66412.html (accessed on 26 June 2020).
56. Abramoff, M.D.; Magelhaes, P.J.; Ram, S.J. Image processing with Image. *J. Biophot. Int.* **2004**, *11*, 36–42.
57. Sánchez-Hermosilla, J.; Medina, R. Adaptive threshold for droplet spot analysis using water-sensitive paper. *Appl. Eng. Agric.* **2011**, *20*, 547–551. [CrossRef]
58. R Core Team. *R: A Language and Environment for Statistical Computing*; R Foundation for Statistical Computing: Vienna, Austria, 2019; Available online: https://www.R-project.org (accessed on 1 May 2020).

Article

Development of Drift-Reducing Spouts For Vineyard Pneumatic Sprayers: Measurement of Droplet Size Spectra Generated and Their Classification

Marco Grella [1,*], Antonio Miranda-Fuentes [2], Paolo Marucco [1], Paolo Balsari [1] and Fabrizio Gioelli [1]

1. Department of Agricultural, Forest and Food Sciences (DiSAFA), University of Turin (UNITO), Largo Paolo Braccini 2, 10095 Grugliasco, Italy; paolo.marucco@unito.it (P.M.); paolo.balsari@unito.it (P.B.); fabrizio.gioelli@unito.it (F.G.)
2. Department of Rural Engineering, University of Cordoba, Edf. Leonardo da Vinci, Ctra. Nacional IV, km 396, 10014 Cordoba, Spain; g62mifua@uco.es

* Correspondence: marco.grella@unito.it; Tel.: +39-011-670-8610

Received: 9 September 2020; Accepted: 2 November 2020; Published: 4 November 2020

Featured Application: The newly designed pneumatic spout generates droplet-sized spectra in a wide range of options, from very fine to very coarse, thus providing farmers with a real possibility for accomplishing the drift-reducing EU policy requirements when using pneumatic sprayers.

Abstract: Pneumatic spraying is especially sensitive to spray drift due to the production of small droplets that can be easily blown away from the treated field by the wind. Two prototypes of environmentally friendly pneumatic spouts were developed. The present work aims to check the effect of the spout modifications on the spray quality, to test the convenience of setting the liquid hose out of the spout in cannon-type and hand-type pneumatic nozzles and its effect on the droplet size, homogeneity and driftability in laboratory conditions. Laboratory trials simulating a real sprayer were conducted to test the influence of the hose insertion position (HP), including conventional (CP), alternative (AP), outer (OP) and extreme (XP), as well as the liquid flow rate (LFR) and the airflow speed (AS) on the droplet size (D50, D10 and D90), homogeneity and driftability (V100). Concurrently, the droplet size spectra obtained by the combination of aforementioned parameters (HP × LFR × AS) in both nozzles were also classified according to the ASABE S572.1. Results showed a marked reduction of AS outside the air spout, which led to droplet size increase. This hypothesis was confirmed by the droplet size spectra measured (D50, D10, D90 and V100). A clear influence of HP was found on every dependent variable, including those related with the droplet size. In both nozzles, the longer the distance to CP, the coarser the sprayed drops. Moreover, LFR and AS significantly increased and reduced droplet size, respectively. A higher heterogeneity in the generated drops was obtained in XP. This position yielded V100 values similar to those of the hydraulic low-drift nozzles, showing an effective drift reduction potential. The classification underlines that the variation of HP, alongside AS and LFR, allowed varying the spray quality from very fine to coarse/very coarse, providing farmers with a wide range of options to match the drift-reducing environmental requirements and the treatment specifications for every spray application.

Keywords: pesticide application; pneumatic nozzle; homogeneity; airflow speed; liquid flow rate; spray drift reducing; spray quality

1. Introduction

Grape production represents one of the most important agricultural businesses worldwide, with 7.5 million hectares and five countries representing 50% of the world vineyard harvested area: Spain

(13%), China (11%), France (10%), Italy (9%) and Turkey (7%) [1]. However, considering the total grape production, Italy ranks second after China, with 8.4 and 13.7 Mt respectively, followed by USA, France and Spain [1]. Considering wine grapes, only the EU ranks first with 61.3% of total world production [1].

Producing high-quality food in a safe manner is one of the most important goals in modern agriculture [2]. Indeed, the high number of spray applications during the vegetative season in intensive production systems determines a massive Plant Protection Product (PPP) use that can cause undesirable effects related to pesticide residues on food and adverse effects to the environment.

At present, PPP spraying is a commonly used technique to control pests and diseases in commercial crops. Nevertheless, spraying is a complex process in which several factors interfere [3]. For that reason, controlling all factors is nearly impossible, and as a result PPP applications could be inefficient. This fact led the European Authorities to develop Directive 2009/128/EC on sustainable use of pesticides [4]. This Directive specifies that pesticide applications should always consider the principles of integrated pest control strategies. Among these, efficiency in the pesticide application necessarily plays a major role that is not completely matched nowadays. Among the main parameters to be considered regarding spraying efficiency, liquid flow rate, airflow rate and droplet size are three of the most important [5–14], along with the forward speed [15,16].

Indeed, during spray applications in bush/tree crops with conventional sprayers, only a limited fraction of the total amount of PPP is deposited on the intended target according to the canopy characteristics [7,17]. Therefore, part of the applied PPP can be transported outside the sprayed area by the action of air currents during the application process as spray drift [18]. Some authors have quantified that, during a spray application, up to 50% of the total applied PPP spray mixture can be lost to the air from the targeted site to a non-target receptor site [19–27]. In addition to the more localized movement of agrochemical residues in turbulent air masses downwind of the application area, residues can also become concentrated in inversions or stable air masses and be transported at long distances [28]. Another fraction could end up as spray deposition on the ground, directly in the field tractor path and underneath target tree rows or indirectly in the adjacent area [5–7,20,29,30]. Among spray losses, spray drift remains the most troubling, as it is really difficult to control [23,31–36], especially in 3D crops such as vineyards [37]. Due to the importance to minimize PPP spray drift generation, strong efforts have been undertaken to properly study this phenomenon and to give adequate advice to farmers around Europe [38].

Among the different factors influencing spray drift, wind speed and direction are the main ones [39]. The higher the wind speed, the higher carrying effect it will have for droplets; therefore, the spray drift risk increases [40,41]. A solution to avoid spray drift is necessary, and it relies on not spraying when the wind is present. Nonetheless, reality is not so simple, as there are many situations in which pesticide needs to be applied a certain day or in a very narrow time window, so the farmer needs to spray even if the wind conditions are not favorable. Consequently, they must act on technical factors (those controlled by the applicator) through a proper adjustment of the spraying equipment [19,42].

Droplet size has proven to be the most effective factor in reducing spray drift [12,20,23–25,28,33,43]. The main reason why this factor affects spray drift is the weight of the emitted drops: the bigger the size, the higher the weight and, therefore, the lower the drift, as the carrying effect of wind will be lower with heavier drops [34]. This was the way followed to develop drift-reducing hydraulic nozzles decades ago. These nozzles produce coarser droplets than the conventional ones do. This result is achieved either by (i) decreasing liquid pressure in the nozzles chamber (drift-guard (DG) nozzles) or (ii) by generating air-inflated drops (air-inclusion (AI) nozzles). As it is the most common strategy to reduce PPP drift, recently different studies have been focused to demonstrate that drift-reducing nozzles generate important spray savings while keeping the necessary spray deposition to ensure the biological efficacy of treatments [44,45]. Nevertheless, some farmers still prefer conventional nozzles as low-size drops generate higher coverage on the leaf surface. As a consequence [46], it is commonly believed they can achieve a higher efficacy against pest and diseases. Several studies have instead showed that droplet size could be increased to reduce spray drift risks without compromising the treatment efficacy [35,45,47], especially in unfavorable environmental conditions [48].

Pneumatic spraying is a well-known technique for its fine droplet size generation. As aforementioned, small droplets are generally requested by the farmers to achieve a uniform target spray coverage that is essential for contact PPP; thus, this technique is very widespread, especially among large-farm vine growers in America and Europe [49,50]. In Italy there are large areas in which vineyard treatments are performed almost exclusively with pneumatic sprayers [51]. The small droplet size produced by this type of nozzle increases the drift risk with respect to the hydraulic nozzles [52]. Thus, whilst the droplet volume median diameter (VMD) generated by conventional hydraulic hollow-cone nozzles typically ranges from 100 to 200 μm, for pneumatic nozzles the VMD is generally below 100 μm [53–55]. This threshold of 100 μm is broadly considered as the minimum diameter that droplets should have to reasonably limit the spray drift risk [56,57]. Therefore, a droplet population with diameters below this value is subjected to losses by drift even with slow environmental air currents.

In order to better understand the droplet population generated by pneumatic nozzles, Balsari et al. [53] investigated and quantified the effect of the main operational parameters, namely the liquid flow rate and the airflow rate, on different variables related to the droplet population, mainly their size and homogeneity. They also assessed the driftability of the drops according to the aforementioned parameters. In order to do so, they developed a laboratory test bench to simulate a real pneumatic sprayer and to test different combinations of parameters, measuring the droplet diameter with a laser device. Once the pneumatic spray behavior was well understood, the same authors planned a strategy to increase droplet size in case of high-speed environmental wind conditions [55]. The idea was to alter the liquid release position inside the air spout, taking advantage of the air speed decrease across the outermost section of the spout [55]. They obtained very promising results, recently confirmed by field trials [58], that demonstrated this strategy could have an important potential to properly reduce spray drift in pneumatic spraying thanks to the droplet size spectra increase.

The main objective of this work was to develop, for the first time, drift-reducing spouts, namely cannon and hand spout types, for vineyard pneumatic sprayers. In detail we evaluated i) the possibility to increase droplets size by adjusting the water income position out of the nozzle spouts and ii) its effect on the droplet size population and homogeneity. Furthermore, the feasibility of introducing alternative positions for the liquid hose in order to achieve different droplet size spectra populations and then different capabilities of drift reductions according to the liquid hose positions were investigated.

2. Materials and Methods

2.1. Trial Location and Spraying Equipment

The laboratory trial was carried out in the facilities of the Department of Agricultural, Forest and Food Sciences (DiSAFA) of the University of Turin located in Grugliasco (Turin, Italy). A test bench was used to simulate the working conditions of the pneumatic sprayer. This test bench consisted of both pneumatic and hydraulic circuits mounted over three different spaces, as shown with detail in Miranda-Fuentes et al. [55]. These spaces included a spraying area, a droplet size measurement area and a control and data acquisition room.

The test bench liquid circuit simulated those mounted in real pneumatic sprayers, including a water tank, a membrane pump AR 202 (Annovi Reverberi S.p.a, Modena, Italy) driven by an electric engine equipped with pressure regulation valves, and a manometer characterized by 0.02 MPa resolution to precisely control the liquid pressure. The test bench ended up in the liquid hose inserted in the air spout of the pneumatic circuit. The liquid flow rate was controlled through a plastic disc with calibrated holes in its perimeter. The air assistance to achieve the spray in the test bench consisted of a centrifugal fan 500 mm diameter (CIMA S.p.a., Pavia, Italy) controlled by a central unit, in which the electric intensity flowing to the driving electric engine, and proportionally the rotary speed of the fan, could be manually adjusted. This control box could set the rotary speed (rev min^{-1}) of the electric engine, measured by a laser tachometer integrated in the system. Thus, even when the control parameter was the amperage, the rotary speed was the indicator to manually regulate the system. As it was previously mentioned, both circuits merged in the air spout, in which the liquid was released through the liquid hose at constant pressure. The air spouts tested included two different types: a cannon-type spout and a hand-type spout (Figure 1a) generally installed on the spray head TC.2M2C.50P (Cima S.p.a., Pavia, Italy). These two kinds of spouts are usually simultaneously employed in the commercial pneumatic sprayers operating in vineyards, as shown in Figure 1b.

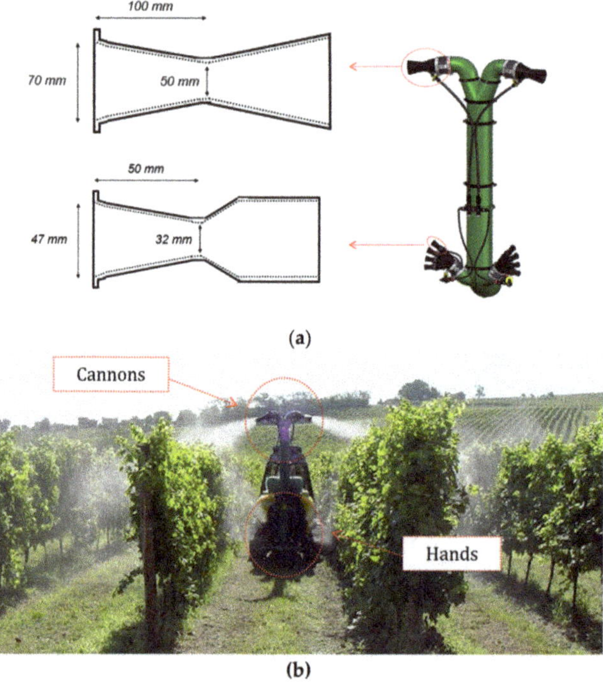

Figure 1. (**a**) Cannon-type and hand-type spouts mounted on the spray head TC.2M2C.50P for (**b**) multiple-row spray application in a vineyard.

These spouts were conveniently modified to alter the insertion position of the liquid hose, as it showed to have a major importance in the generated droplet size in cannon spouts [55]. The insertion position was, consequently, established as the main variable of the study, and it was called "hose position" (HP). It presented four possible levels in each spout type, called "conventional position" (CP), "alternative position" (AP), "out position" (OP) and "extreme position" (XP), as shown in Figure 2a. The CP position corresponds to that established by the manufacturer for the commercial equipment (Figure 2b,c). The other hose positions, namely AP, OP and XP, varied according to the spout type, as shown in Figure 2b for the cannon and in Figure 2c for the hand spouts. In particular, in both spout-types, the AP corresponds with the one in which the outermost part of the hose is placed coincident with the border of the air spout (Figure 2b,c), and the XP was experimentally set as the maximum distance in which a uniform liquid atomization could be achieved. This distance value was experimentally found by moving the liquid hose out of the air spout along its longitudinal axle through a rail in steps of 5 mm and the droplet size spectra measured in each position. Once this maximum distance was exceeded, the spray cloud changed drastically, appearing as extremely coarse droplets and losing the normal distribution. The OP was set at an intermediate distance between the AP and the XP (Figure 2b,c).

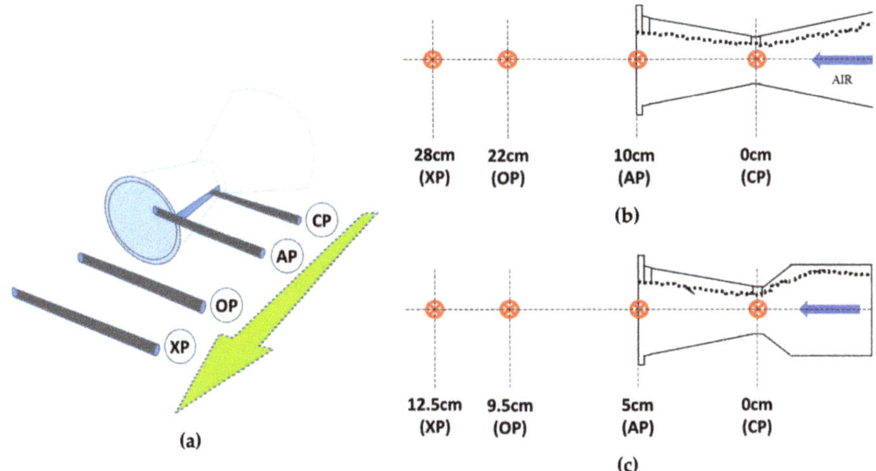

Figure 2. (**a**) Liquid hose positions from the inner to the outer part of spouts (green arrow): conventional position (CP) (0 cm reference) passing through alternative position (AP), out position (OP) and extreme position (XP). Distances (cm) from the reference position were different for (**b**) the cannon-type and (**c**) the hand-type spouts.

2.2. Droplet Size Spectra and Airflow Speed Measurements

The droplet size spectra were measured with a Malvern Spraytec® laser diffraction system STP5342 (Malvern Instruments Ltd., Worcestershire, UK) (Figure 3a). The instrument has a maximum measurement frequency of 10 kHz and a measurement range of 0 to 2000 μm. The instrument includes software (SprayTec Software v3.30, Malvern) for managing the data acquisition and charting. This software directly acquired the droplet size parameters D50 (or Volumetric Mean Diameter, VMD), D10 and D90. It also calculated V100, the fraction of the spray with droplets below 100 μm in diameter, which easily can be blown away by the wind according to different authors [56,57]. The droplet homogeneity could also be drawn from the aforementioned droplet size parameters, if expressed as the RSPAN Factor (RSF), which can be calculated as shown in Equation (1) [12,59].

$$RSF = \frac{D90 - D10}{D50} \tag{1}$$

where RSF is dimensionless, and D90, D10 and D50 are expressed in μm.

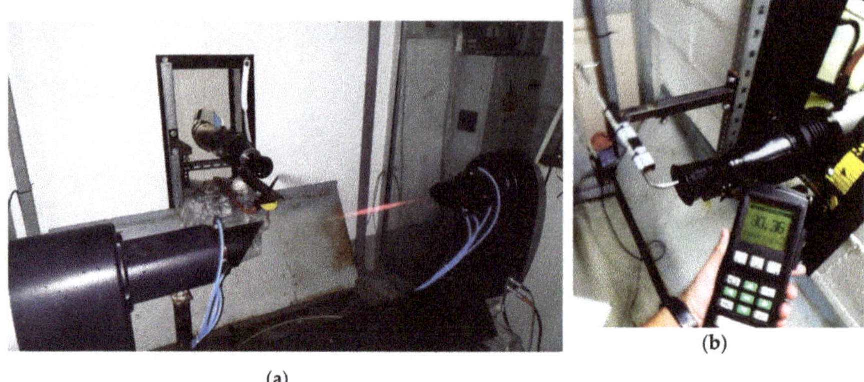

(a) (b)

Figure 3. (**a**) Malvern Spraytech® laser diffraction system STP5342 for the measurements of droplet size spectra and (**b**) Testo 400 Pitot-tube-based anemometer for the measurements of airflow speed generated by both cannon-type and hand-type spouts.

The conventional hydraulic hollow cone nozzle Albuz® ATR lilac and the air-induction Albuz® TVI 8001 (CoorsTek Inc., Evereux, France) both operated at 0.7 MPa pressure, were selected as reference conventional spray technology for the comparison with the droplet size spectra investigated using the modified pneumatic spouts. In particular, the ATR lilac nozzle is well known to produce very fine (VF) spray quality, likewise the TVI 8001 nozzle is well known as a drift-reducing nozzle for the ultra-coarse (UC) spray quality generated [43]. Furthermore, the Albuz® ATR lilac was studied by other authors, and it is widely used by farmers with conventional hydraulic air-assisted sprayers both in vineyards and orchards [60]. The droplet size spectra generated by the reference nozzle were measured using the same laser diffraction system device described above and the methodology usually applied for the measurements of droplet size spectra generated by hydraulic nozzles, fully detailed in Grella et al. [25].

The measurement of the airflow speed was performed with a Pitot-tube-based anemometer Testo 400 (Testo SE & Co. KGaA, Lenzkirch, Germany) (Figure 3b) at 1Hz frequency with a measurement resolution of ±0.01 hPa, a differential pressure corresponding to 1.28 m s^{-1}, and a measurement range of up to +2000 hPa (571.43 m s^{-1}). The instrument was fixed to an ad hoc support developed by the researchers with some modifications to properly adapt it to the experiment [55]. The main advantage of this support was the possibility to keep the air speed measurement instrument in the center of the air spout (Figure 3b), avoiding errors that could deeply affect the measurements when holding it manually. The modifications in the instrument were essentially aimed at an enlargement of the holding tube to measure speed in the outermost position of the liquid hose (XP). Thus, the airflow speeds were measured along the central axis in six positions, corresponding to 0, 5, 10, 16, 22 and 28 cm to the CP for the cannon-type spout and in five positions, corresponding to 0, 2.5, 5.0, 9.5 and 12.5 cm in a single spout for the hand-type nozzle. Preliminary trials were conducted to ensure the airflows generated individually by the four spouts that composed the hand-type nozzle were comparable. In each sampling position, the data were recorded for 60 s per each of the three replicates performed.

2.3. Spray Parameters, Experimental Design and Settings Used During Trials

Three different liquid flow rates (LFRs) were investigated, namely 1.00, 1.64 and 2.67 L min^{-1}. The intended LFR were obtained setting the liquid circuit pressure at 0.1 MPa and using the liquid flow regulator disc (a plastic disc with calibrated holes in its perimeter) in positions 3, 5 and 7, respectively (Table 1). As mentioned above, four HPs were tested in both pneumatic nozzle types, namely CP, AP, OP and XP (Table 1). The intended exact positions of liquid hose, relative to the spout, were guaranteed thanks to a support integral with the spout body (Figure 3a). Concurrently, four AS were tested, namely

81.3, 90.0, 100.2 and 109.2 m s^{-1} for the cannon-type nozzle and 57.9, 64.6, 74.3 and 84.2 m s^{-1} for the hand-type nozzle. The tested AS values were referred to the reference position (CP), since the increase in spout section diameter deeply affects the decrease of AS along the spout body [55]. The intended AS were obtained properly setting the test bench fan rotary speed to simulate real conditions during spray application. Therefore, the main complexity was to adjust the test bench to make sure it matched the working parameters present in a real sprayer. There are many pneumatic circuit parameters that could be compared in both the bench and the sprayer, but the most representative and the one that ensures an equivalency in the working conditions is the air pressure. Thus, air pressure measurements were done to adjust both systems for every kind of pneumatic nozzle. The fully explained details of the calibration for both cannon- and hand-type nozzles along with the calibration results can be found in Balsari et al. [53] and Miranda-Fuentes et al. [55]. The air pressure measurements were performed with a manometer with a measurement resolution of ±1 mm H_2O, which was attached to a plastic piece fixed to the nozzle in every case. These pieces were large enough to reach the central part of both air spouts. The air pressure was measured for different Power Take Off (PTO) speed values in order to match the intended AS values previously measured in the real sprayer. According to those measurements, for laboratory trials the test bench fan rotary speeds were set at 541, 598, 663 and 720 rev min^{-1} for the cannon-type nozzle and at 488, 536, 609 and 677 rev min^{-1} for the hand-type nozzle, respectively. The operating parameters assessed are listed in Table 1.

Table 1. Operating parameter values of the two nozzle types used in laboratory trials.

	Parameter *	Levels	Regulation Basis	Test Settings
Cannon nozzle	LFR	1.00/1.64/2.67 L min^{-1}	Flow regulator disc	Positions 3/5/7
	AS	81.3/90.0/100.2/109.2 m s^{-1}	Fan rotary speed	541/598/663/720 rev min^{-1}
	HP	0/10/22/28 cm **	Hose support	CP/AP/OP/XP ***
Hand nozzle	LFR	0.84/1.33/2.07 L min^{-1}	Flow regulator disc	Positions 3/5/7
	AS	57.9/64.6/74.3/84.2 m s^{-1}	Fan rotary speed	488/536/609/677 rev min^{-1}
	HP	0/5/9.5/12.5 cm **	Hose support	CP/AP/OP/XP ***

* LFR: liquid flow rate, AS: Air speed, HP: liquid hose position. ** Distance from the conventional position (CP). *** CP: conventional position, AP: alternative position, OP: out position, XP: extreme position.

The experimental design was completely randomized, as no restrictions were present when measuring the droplet parameters (D50, D10, D90, RSF and V100) for a given combination of values of the independent variables (HP, LFR and AS; Table 1) in each studied pneumatic nozzle. The treatment order was randomized to avoid possible influence of external factors not considered in the design. The dependent variables were the droplet size parameters (i.e., D50, D10 and D90), the droplet homogeneity, given by the RSF, and the droplet driftability, given by the V100 parameter.

The trial began with the air spout placement, and it was conducted in two steps. In the first one, the airflow rate was established by adjusting the fan's rotary speed to the values obtained in the calibration [53,55]. The air speed was then measured in different positions using the Pitot tube, as described before. In the second step, the spray was enabled, and a 30 s time was given to the system to stabilize. Once the time had passed, the laser diffraction instrument was settled at the most appropriate distance from the spout (50 cm in accordance to previous trials [55]), and a total of 30 measurements were taken at 1 Hz acquisition frequency. The operational parameters (Table 1) were then combined to test every possible combination. In total, 48 configurations were tested for each spout type, deriving from the combination of different LFR, AS and HP. Three test replicates for each spout configuration were done.

2.4. Data Analysis

The data were recorded by Malvern Spraytech® software, while a special macro was developed in R-Studio [61] to automatically import and compile the data of every replication with their work parameter combination labels. Data analysis was performed in IBM SPSS Statistics v25 [62]. Normality and homoscedasticity of data were checked by using Shapiro–Wilk and Levene tests (α=0.05 in both cases). Graphs were done with the SPSS and Excel graph editor.

A three-way ANOVA test (α=0.05) was run to test the influence of the three factors (HP, LFR and AS) on every single dependent parameter (D50, D10, D90, RSF and V100). A FREGW (Ryan-Einot-Gabriel-Welsh F) test (α=0.05) was then applied to set the homogeneous groups. Correlations were developed between D50 and V100 for both pneumatic nozzles tested.

The cumulative sprayed volume curves, obtained by each tested configuration together with the reference hydraulic nozzles were compared with American Society of Agricultural and Biological Engineers (ASABE) nozzle classifications (ASABE S572.1) [63].

3. Results and Discussion

3.1. Airflow Speed Drop Along the Spouts

The airflow speeds measured along the longitudinal axle of cannon and hand spouts are graphed separately in Figure 4. Specifically, Figure 4a shows an important decrease in the air speed from the inner to the outermost position for the cannon-type nozzle (about 55% in every case). Nevertheless, this decrease was not constant along the axle, registering a slight slope between 5 and 10 cm from the inner position and very important slopes from 16 cm on (Figure 4a). Thus, the speed decreased at a variable rate ranging from 0 to 3.67 m s^{-1} cm^{-1}. The effect of the nozzle border on the speed decrease was noticeable: the air speed decrease was less extreme near of the spout border, becoming more important in the 12 cm after the first 6 cm interval. This effect could be due to a border effect of air enclosure before spreading out of the main current direction in the following sections. Concurrently, Figure 4b shows the response of the air speed in the hand-type nozzle. The two first sections, between the OP and the AP, kept the air speed relatively constant, with less than 10% decrease in total in the most extreme case (677 rev min^{-1} of the test bench, Figure 4b). After the air spout border, there was an abrupt decrease in the air speed, with a maximum mean value of 7.56 m s^{-1} cm^{-1} in the case of the maximum test bench fan rotary speed.

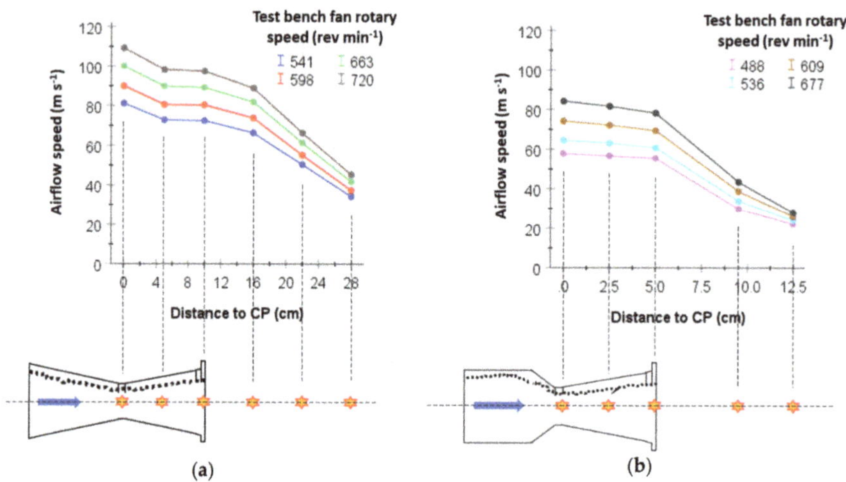

Figure 4. Airflow speed (m s^{-1}) measured along the longitudinal axle of the spout at different distances (cm) from the conventional position of liquid hose (CP) at four test bench fan rotary speeds: (**a**) cannon-type and (**b**) hand-type pneumatic nozzles. Error bars show the ± Standard Error of the Mean.

3.2. Droplets Size Spectra Measured

The VMDs, or D50 values, distribution per liquid hose positions are shown separately by nozzle type in Figure 5. In general, the cannon (Figure 5a) generated finer droplets than the hand-type nozzle did (Figure 5b), with median values in their conventional insertion position of the liquid hose (CP) of 76 vs 95 μm, respectively. This fact is consistent with the observations made in previous droplet size characterization tests [53]. Nonetheless, there was a higher capacity to proportionally increase D50 values in the cannon (265% vs 223% total increase from CP to XP, Figure 5), which makes it possible to produce relatively similar drops in the XP in both the cannon- and hand-type nozzles (270 vs 307 μm, Figure 5). When looking at the distribution of the D50 droplets in the different tested positions, values in the cannon spout were more homogeneously distributed than those in the hand spout. Thus, the first nozzle generated a D50 value of 74 μm at the reference position (CP), whereas values were 112, 205 and 270 μm at AP, OP and XP positions, respectively. The last three values corresponded to 51%, 177% and 265% droplet diameter increase when compared to the reference position (CP). For the hand-shaped nozzle, a 95 μm droplet diameter was recorded at CP. Diameters at AP (145 μm), OP (160 μm) and XP (307 μm) were 53%, 68% and 223% larger, respectively, than those recorded at CP. As it can be seen (Figure 5), values were more heterogeneous in the hand-type nozzle (b) when compared to the cannon ones (a). The D50 variability (shown by the interquartile rate in Figure 5) in both nozzles increased with increasing distances from CP. This is also consistent with previous works characterizing pneumatic sprayer droplet size, as the droplet size indicators (D50 among them) can be predicted with linear models, which are affected by the liquid flow rate and air speed [53]. When comparing these global results with the air speed decrease rate (Figure 5), data were consistent in the cannon-type nozzle. Air speed decrease did not present extreme values along any of the measured sections, so a progressive droplet size increase was expected. On the other hand, the hand-type nozzle presented more extreme air speed decrease values, so the higher heterogeneity of data would be explained.

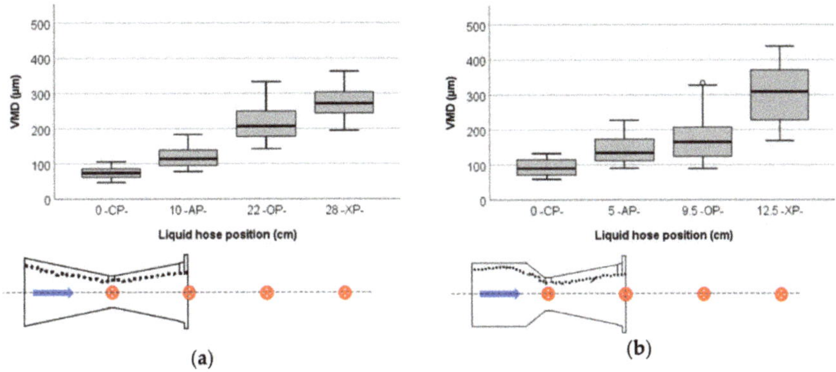

Figure 5. Volume Median Diameter (VMD) (μm) per liquid hose position (HP): (**a**) cannon-type and (**b**) hand-type spouts. CP, conventional insertion of liquid hose position; AP, alternative position; OP, out position; XP, extreme position.

3.2.1. Cannon-Type Nozzle

D50, D10 and D90 values were significantly affected by the liquid hose position (HP), the LFR and the AS ($p < 0.001$, Figure 6a). In general, an increment in the droplet size can be observed for every indicator (D50, D10 and D90) with the increase of the liquid hose insertion distance to the original one. In this sense, moving the hose out of the spout generated an important increment in the droplet size, as expected by taking into account the air speed decrease found in the present work (Figure 4). This fact was also summarized in Figure 5a, where a clear increase in this parameter was found when increasing distance. As shown in previous works, air speed and liquid flow rate play key roles on droplet size [53,55]. Thus, the first parameter generates a clear decrease on the drop size and the second one behaves the opposite. Regarding D50, nearly every value obtained in the CP was lower than 100 μm (Figure 6a). This contrasts with the case of the XP, where every value was above 200 μm. AP and OP presented intermediate responses. Maximum and minimum D50 mean values were 325 μm (with XP, minimum LFR and minimum AS) and 49 μm (with CP, minimum LFR and maximum AS). In general, increments of D50 values were in the range 30%–94%: between 50% and 75% moving the liquid hose position from CP to AP, between 79% and 94% when the liquid hose was shifted from AP to OP and between 30%–43% when it was further moved to XP.

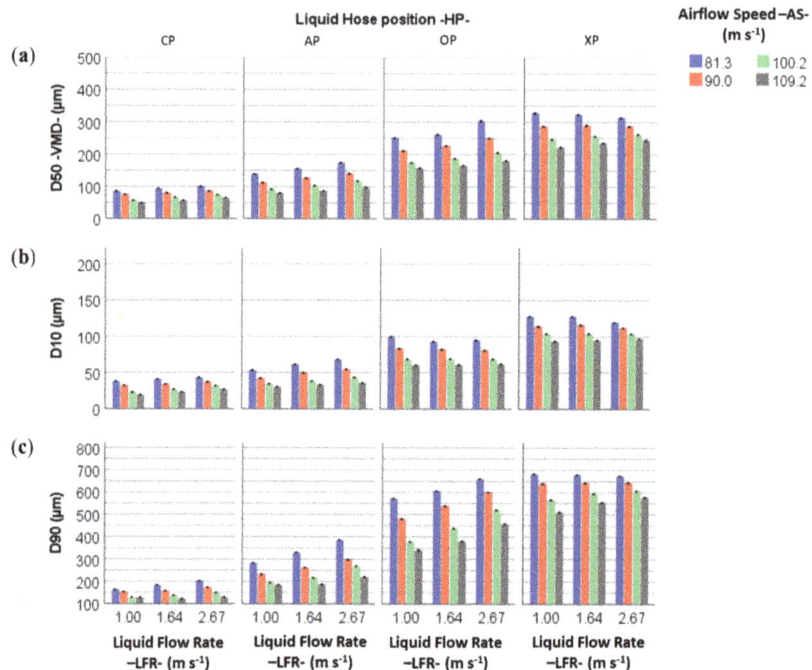

Figure 6. (**a**) D50 (Volume Median Diameter, VMD), (**b**) D10 and (**c**) D90 values (μm) per liquid hose position (HP), liquid flow rate (LFR) and airflow speed (AS) for the cannon-type nozzle. Bars show the mean ±Standard Error. CP, conventional insertion of liquid hose position; AP, alternative position; OP, out position; XP, extreme position.

D10 values for CP resulted below 45 μm whilst the same values for XP were all above 90 μm (Figure 6b). This fact has important practical implications because the use of pneumatic nozzle with the liquid hose in extreme position, allows to generate a very important fraction of the spray relatively safe from drifting, as it will be discussed in the V100 section. The maximum and minimum D10 mean values were 126 μm (in XP combined with minimum LFR and minimum AS) and 17 μm (in CP combined with minimum LFR and maximum AS; Figure 6b). The D10 mean increase from CP to AP ranged, in this case, from 36% to 80%. The increase range between AP and OP was 52% to 89%, whilst the same for OP to XP was 27% to 37%. In this case, the atypical response of D10 decrease with the LFR increase was also found in the OP (Figure 6b).

D90 values comprised a minimum value of 128 μm (in CP combined with minimum LFR and maximum AS) and a maximum of 682 μm (in XP combined with minimum LFR and minimum AS; Figure 6c). The CP resulted, in general, in values below 200 μm, whilst the XP yielded values above 500 μm. The mean D90 increase values were 73% from CP to AP, 101% from AP to OP and 18% from

OP to XP. In the same way as D50 and D10, D90 values showed their highest heterogeneity in XP with relation to the rest of positions.

3.2.2. Hand-Type Nozzle

Similarly to the cannon-type nozzle, the three studied droplet size variables in the hand-type nozzle also depend on the HP, LFR and AS ($p < 0.001$). This fact is consistent with the air speed decrease found in the hand-type nozzle (Figure 4). General droplet size interval for every tested position can be observed in Figure 5b. Droplet size was more heterogeneous in the XP compared to that observed in the cannon spout.

The D50 values measured for the hand-type nozzle (Figure 7a) were slightly higher than those obtained in the cannon spout. Thus, CP resulted in D50 below 150 µm, with the minimum mean value of 62 µm for the combination between minimum LFR and maximum AS. XP, on the other hand, resulted in D50 values above 125 µm, with a maximum value of 407 µm for the combination of minimum LFR and minimum AS. In this particular case, the atypical behavior already observed for the cannon was also registered in the hand-type spout, with a decrease in the droplet size with the increase in the LFR. There was also a considerable influence of the other two evaluated parameters ($p < 10^{-3}$) on D50. In the case of LFR, there was a positive effect on D50, whilst the opposite was found for AS, as expected from previous works [53]. The mean D50 increase that can be found between CP and AP was 42%. From AP to OP there was a mean increase of 39%. Last, OP to XP had a mean VMD increment of 75%.

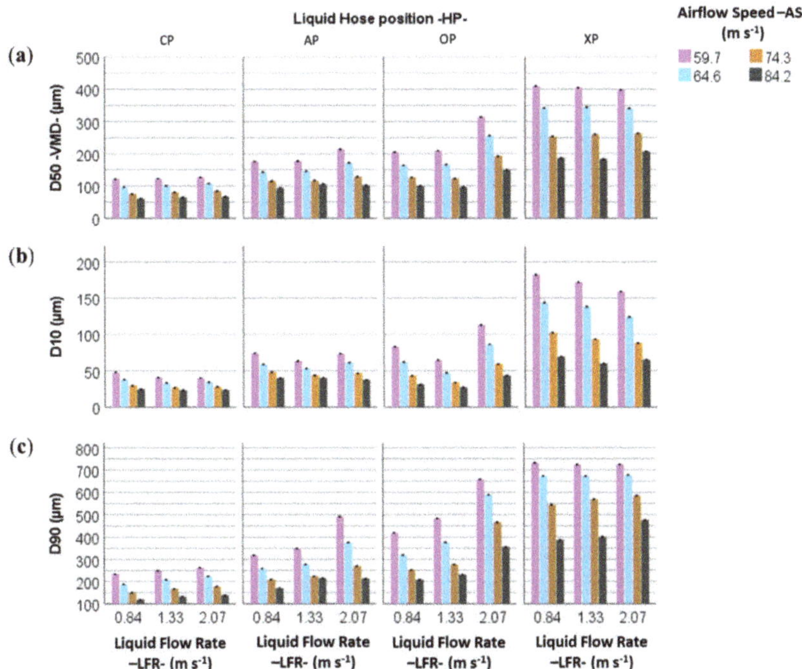

Figure 7. (**a**) D50 (Volume Median Diameter, VMD), (**b**) D10 and (**c**) D90 values (µm) per liquid hose position (HP), liquid flow rate (LFR) and airflow speed (AS) for the hand-type nozzle. Bars show the mean ± Standard Error. CP, conventional insertion of liquid hose position; AP, alternative position; OP, out position; XP, extreme position.

The D10 values (Figure 7b) comprised a minimum value of 23 µm for the combination of CP, intermediate LFR and maximum AS, and 179 µm for the combination of XP, minimum LFR and minimum AS. The CP presented D10 values all below 50 μm, while the XP presented values above 55 μm. The mean increase in this parameter from CP to AP was 49%, from AP to OP it was 3% and from OP to XP it was 135%.

The D90 values (Figure 7c) were all found below 280 μm in CP and above 370 μm in XP. The most extreme values were 112 μm (for CP/min LFR/max AS) and 735 μm (for XP/min LFR/min AS). The mean increase from CP to AP was 35%, from AP to OP it was 28% and from OP to XP it was 94%.

In general, it can be observed that the most extreme positions generated an atypical response of the LFR on the droplet size parameters. Nevertheless, the three LFR levels did not produce great differences in comparison with the AS ones. It could be stated that LFR had a lesser impact on droplet size in the outermost positions of the liquid hose.

3.3. Cumulative Sprayed Volume Curves and Their Classification According to ASABE S572.1

The cumulative sprayed volume curves, obtained with cannon (Figure 8) and hand-type (Figure 9) nozzles by testing all configurations (combination of HP, LFR and AS), compared with the American Society of Agricultural and Biological Engineers (ASABE) nozzles classifications (ASABE S572.1) [63], showed that an appropriate selection of pneumatic sprayer operational parameters, namely LFR and AS, led to a limited change in the spray quality generated, varying from very fine (VF) to fine (F) in the best cases. On the contrary, changing HP deeply affected the spray quality and reached the coarse (C) (Figure 8) and very coarse (VC) (Figure 9) spray quality when the liquid hose was tested in the extreme position (XP) combined with reduced AS, irrespective of LFR. Even if Balsari et al. [53] recognize the importance of a proper selection of LFR and AS parameters in order to vary the spray quality in conventional pneumatic sprayers, likewise it was clear that the spray quality extent change from VF to F was not enough to guarantee environmental safeguard, despite the increased droplet size dimension. For this reason, according to the preliminary tests performed by Miranda-Fuentes et al. [55], the HP change along the pneumatic spouts is proven as the driving factor for changing the spray quality. This strategy allows to achieve a range of different spray qualities comparable to that reachable with hydraulic nozzles, which can be varied in type (conventional vs. air induction) and size [43]. Thus, at the time of application, in both pneumatic nozzle types the selection of HP allowed to vary the spray quality in a wide range without changing the other operational parameters selected for the application (LFR and AS). In detail, the cumulative sprayed volume curves derived from the tested configurations using the liquid hose in conventional position (CP) showed a droplet spectrum similar to the hydraulic nozzles Albuz® ATR lilac used at 0.7 MPa (Figures 8 and 9) and selected as reference. Only the hand-type nozzle tested in CP position with AS of 59.7 and 64.6 m s^{-1} showed a coarser spray quality, namely fine (F), irrespective of LFR (Figure 9). On the contrary, none of the tested configurations with both cannon- and hand-type pneumatic nozzles in XP position achieved the extra coarse (XC) spray quality, likewise that generated by the air-induction Albuz® TVI8001 nozzles (0.7 MPa pressure) used as reference nozzle for the coarse spray quality (Figures 8 and 9). In general, the liquid hose position in AP and OP allowed intermediate spray quality, varying from F to M according to the LFR and AS selected. Even when the hand-type nozzle generated averagely larger droplets than the cannon-type did, the spray quality produced by both of them, for the same tested configuration, was very similar (Figures 8 and 9). In practice, no substantial difference in prospective coverage of the leaves and fruits could be noticed between the two nozzle types. In any case, the possibility to vary in a wide range the droplet size by simply adjusting the liquid hose position gives, for the first time, the opportunity to both farmers and technicians to match the environmental requirements, balancing at the same time the treatment specifications for every spray application while using pneumatic spraying in vineyards. Recently, Grella et al. [58] demonstrated through field trials that the variation in the spray quality over the range investigated (from CP to XP) did not affect the canopy coverage, while coarser sprays (liquid hose in AP and XP) produced greater deposits on the target. Concurrently, the use of cannon spout in XP position significantly reduced the off-field ground losses in the downwind area.

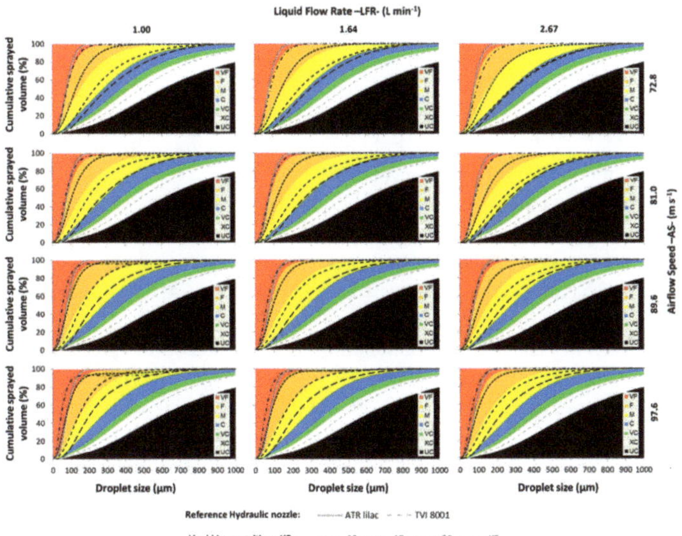

Figure 8. Cumulate sprayed volume (%) curves as a function of droplet size (µm) per liquid flow rate (LFR), airflow speed (AS) and liquid hose position (HP) for the cannon-type nozzle. In each graph the hydraulic reference nozzles hollow cone conventional ATR lilac and air-induction TVI 8001 (Albuz®) are displayed, both operated at 0.7 MPa pressure. CP, conventional insertion of liquid hose position; AP, alternative position; OP, out position; XP, extreme position. VF, very fine; F, fine; M, medium; C, coarse; VC, very coarse; XC, extremely coarse; UC, ultra-coarse/unclassified; (ASABE S572.1).

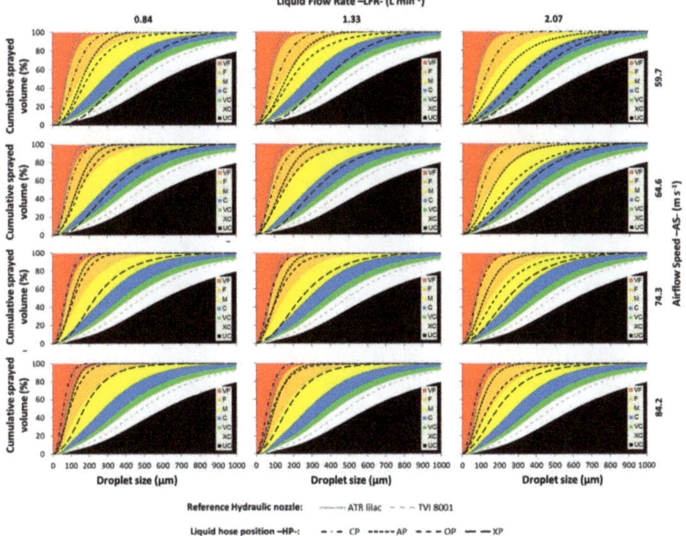

Figure 9. Cumulate sprayed volume (%) curves as a function of droplet size (µm) per liquid flow rate (LFR), airflow speed (AS) and liquid hose position (HP) for the hand-type nozzle. In each graph the hydraulic reference nozzles hollow cone conventional ATR lilac and air-induction TVI 8001 (Albuz®) are displayed, both operated at 0.7 MPa pressure. CP, conventional insertion of liquid hose position; AP, alternative position; OP, out position; XP, extreme position. VF, very fine; F, fine; M, medium; C, coarse; VC, very coarse; XC, extremely coarse; UC, ultra-coarse/unclassified; (ASABE S572.1).

3.4. Droplet Homogeneity

Significant differences ($p < 0.001$) were found in both nozzles for every single variable and their interactions. In general, it could be stated that the distance increase to the CP generates an increase in the droplet heterogeneity (Figure 10). This result is in line with the results obtained in previous works for the CP and the AP [55]. RSF values were slightly lower in the hand spout than in the cannon one (1.63 vs 1.78), especially in the surroundings of the CP.

In the cannon spout, the hose position change, although statistically significant, did not generate major differences in the mean RSF values (Figure 10a). Thus, values for every single position were near 2.00. The observed trend was a reduction in the droplet population homogeneity with the distance increase from the CP. The increase in AS also increased the RSF in general, especially in the case of the lower LFR in the CP (Figure 10a).

There was a much higher influence of the hose insertion position in the case of the hand spout (Figure 10b). In this case all values were around 2.00, but with important differences among the tested hose positions. There was a significant particularity in the hand spout case: values presented a higher RSF in the OP than in the XP, while the opposite would have been the most predictable scenario (Figure 10b). It should be pointed out that in this nozzle both LFR and AS had a higher impact on the RSF than they had in the cannon spout.

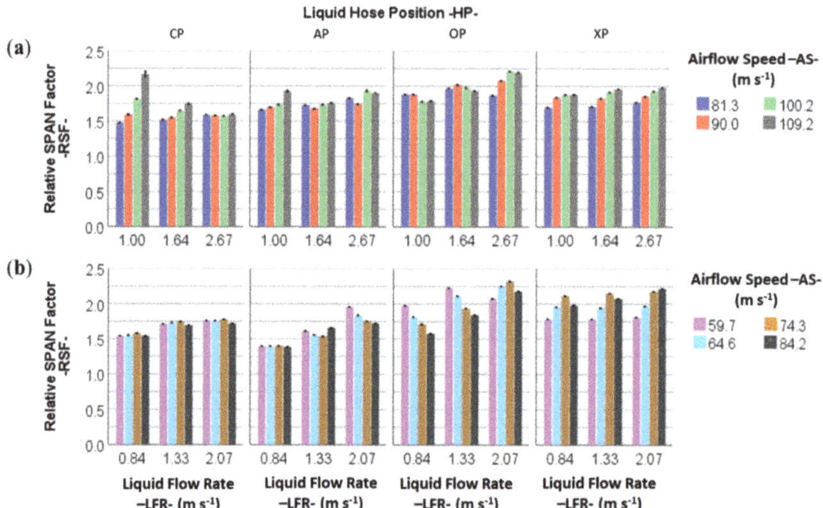

Figure 10. Relative SPAN Factor (RSF) per liquid hose position (HP), liquid flow rate (LFR) and airflow speed (AS) for both the (**a**) cannon-type and (**b**) hand-type nozzles. Bars show the mean ±Standard Error. CP, conventional insertion of liquid hose position; AP, alternative position; OP, out position; XP, extreme position.

The comparison between both nozzles regarding the droplet homogeneity follows a trend similar to that observed in previous works [53]. In those, a positive influence of both LFR and AS was observed on the RSF in both nozzles. Similarly to the present findings, the cannon spout had a higher RSF than the hand-type one did. A marked increase in the droplet heterogeneity was also observed in the maximum AS and minimum LFR combination (Figure 10a). This could be explained by the fact that with a very low LFR the amount of sprayed liquid might be insufficient for the high air current to produce an optimal water division into homogeneous droplets, thus increasing the amount of very fine ones, altering their normal distribution and increasing the droplets heterogeneity by affecting the kurtosis.

3.5. Droplet Driftability

According to other authors [13,43,56,57,64], the V100 is a valuable parameter to predict spray drift. The lower this parameter, the lower the portion of spray droplets below 100 μm and, therefore, the predicted incidence of the spray drift. As it is displayed in Figure 11, V100 values were generally higher for the cannon-type nozzle in the CP and AP. This trend enhanced the spray drift generation during field spray application, as the cannon spout must spray at a longer distance from the target than the hand one in the conventional setting used in farms (Figure 1b). Nevertheless, the present study showed a different behavior of the two nozzle types in the droplet size spectra increase after the AP (liquid hose out of the spout body). Thus, for OP and XP, cannon spout presented lower V100 values when compared to the hand-type one (Figure 11). This finding might be relevant for the spray drift reduction, as it could indicate a higher interest in implementing OP and XP in the cannon spout.

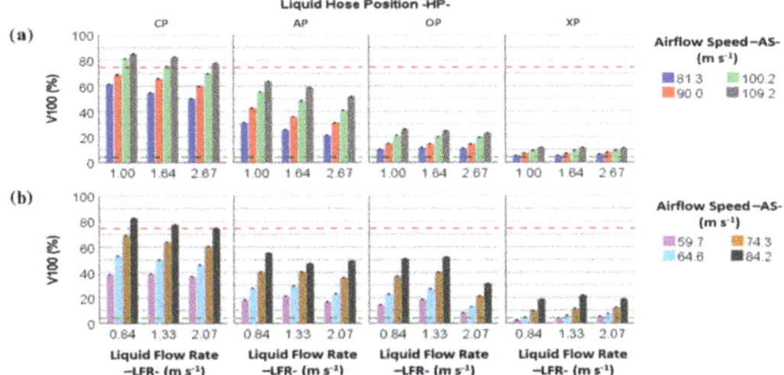

Figure 11. V100 (%) per liquid hose position (HP), liquid flow rate (LFR) and airflow speed (AS) for both the (**a**) cannon-type and (**b**) hand-type nozzles. Bars show the mean ± Standard Error. The dashed lines represent the hydraulic reference nozzles, namely hollow cone conventional ATR lilac in red and air induction TVI 8001 in green (Albuz®), both operated at 0.7 MPa pressure. CP, conventional insertion of liquid hose position; AP, alternative position; OP, out position; XP, extreme position.

Indeed, in the case of the cannon-type nozzle, there was a consistent V100 reduction among consecutive hose positions. Thus, from a mean value of 68% in CP, the V100 mean value decreased to 8% in XP (Figure 11a). Within each liquid hose position, the combination of minimum LFR with maximum AS and maximum LFR with minimum AS deeply affected the driftability. These settings yielded mean V100 values of 49 ± 2% (max LFR/ min AS) and 61 ± 2% (min LFR/max AS) in CP (Figure 10a). AP gave V100 results between 21 ± 1% (max LFR/ min AS) and 64 ± 2% (min LFR/max AS). OP resulted in a range from 10 ± 1% (max LFR/ min AS) to 21 ± 1% (min LFR/max AS). Finally, XP values ranged from 5 ± 1% (min LFR/ min AS) to 10 ± 1% (min LFR/max AS) (Figure 10a). In a practical way, the best V100 reduction within each hose position was always achieved when the system was operated with the maximum LFR and minimum AS, confirming the marked influence of these operational parameters [53].

Comparing V100 values generated by pneumatic nozzles with those generated by the reference hydraulic nozzle (Figure 11a) Albuz® ATR lilac operated at 0.7 MPa pressure (dashed red line corresponding to V100 equal to 74%), it can be observed that similar values were obtained when the pneumatic nozzles were used in CP and combined with the highest AS (89.6 and 97.7 m s^{-1}). Changing the hose position from AP to XP, all possible combinations of tested parameters reduced the V100 level compared with the reference nozzle. This finding reflects the capability of AP, OP and XP to generate droplet populations with a driftability similar to those produced by a wide range of conventional hydraulic nozzles characterized by different dimensions and operated in a wide range of liquid pressure. Specifically, the V100 values for XP were similar to those obtained using hydraulic air induction nozzles. Indeed, the cannon-type nozzle operated in the XP position and, combined with the lowest AS, achieved V100 values fully comparable with those obtained with air induction nozzle Albuz® TVI8001 (V100 equal to 4%) marked with the dashed green line in Figure 11a. The minimum value achieved in the XP was 5%, so the droplet size can be considered similar to that of a conventional low-drift nozzle. Considering the configurations characterized by the combination of highest AS (97.6 m s^{-1}) and highest LFR (2.67 L min^{-1}), the switch of HP from CP to XP determined a reduction of V100 equal to 90%, a value fully confirmed by field trials that showed a 95% reduction in off-target losses [58].

Regarding the hand-type spout, there was a significant reduction in V100 across the different hose positions, turning from a mean value of 58% in CP to 9% in XP (Figure 11b). Nevertheless, there was a low reduction between AP and OP. Indeed, the mean V100 value only decreased from 29% to 28%. The lowest possible value in the conventional nozzle configuration was 37 ± 1% for the maximum LFR with the minimum AS (Figure 11b). AP originated values between a maximum mean value of 53 ± 3% (med LFR/max AS) and a minimum of 8 ± 2%. OP yielded values between 56 ±1% (min LFR/max AS) and 16 ± 1% (max LFR/min AS). Finally, XP resulted in values from 22 ± 1% (med LFR/max AS) and 3 ± 1% (min LFR/min AS) (Figure 11b). Comparing these values with the reference nozzle Albuz® ATR lilac operated at 0.7 MPa pressure (V100 equal to 74%), only the liquid hose in CP combined with the highest AS generated values higher than the threshold marked with a dashed red line in Figure 11b. Concurrently, the liquid hose in XP combined with the lowest AS was able to lower the V100 threshold fixed by the reference air induction nozzle Albuz® TVI8001 marked with the dashed green line in Figure 11b. The very low V100 values achieved in the last position of the liquid hose were even lower than many of the ones reported for the hydraulic low-drift nozzles [43]. Even if the V100 values measured in this study (74%) for the Albuz® ATR lilac operated at 0.7 MPa pressure were higher than those obtained by Zande et al. [43] (equal to 23%), the droplet size spectra parameters were fully in accordance with those reported by ASABE S572.1 classification, and the nozzle falls in the same class, namely very fine (VF) [63].

As it can be drawn from Figure 12, there was a clear correlation between V100 and D50 in both cases ($p < 0.001$). Determination coefficients were very similar (0.9764 and 0.9704 in the cannon-type (Figure 12a) and hand-type (Figure 12b) nozzles), and the correlation coefficients were nearly the same (−0.01 and −0.009). This result indicates that both nozzles have similar behavior. Moreover, this kind of response can be expected from different pneumatic nozzles. In the case of the cannon, there was a deviation of the regression line with high D50 values. This behavior can be expected when taking into account the higher variability that is found in the most extreme positions of the liquid hose (Figure 12a). A similar trend can be noticed in the hand nozzle for which a loss of fitting in the lowest D50 values can be observed (Figure 12b).

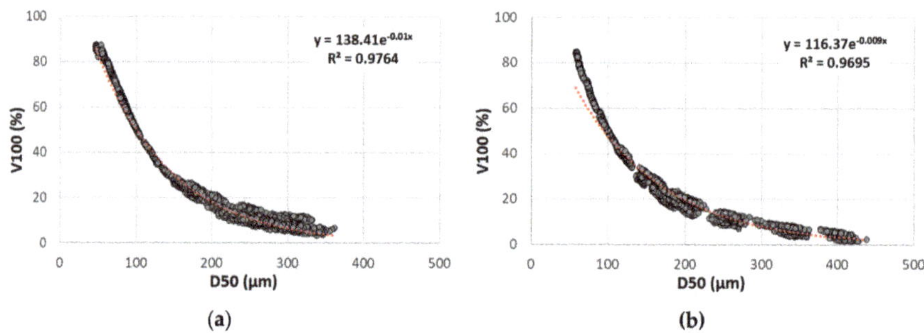

Figure 12. Correlation between D50 (μm) and V100 (%) for all tested configurations (combination of liquid hose insertion position (HP), liquid flow rate (LFR) and air speed (AS) for both the (**a**) cannon-type and (**b**) hand-type nozzles.

4. Conclusions

Following previous findings related to the droplet size increase in pneumatic spraying, a study was carried out to check the influence of modifications in the liquid hose position with respect to the air spout. In both tested nozzles, namely cannon and hand types, the air speed decreased consistently from the inner to the outer part of the spout, with higher decrease outside of the spout.

Therefore, the strategy to move the liquid release hose to outer positions with respect to the conventional one is as an effective way to substantially increase the droplet size spectra generated by pneumatic sprayers. This effect was certified by the Volume Median Diameter (D50) increase equal to 280% for the cannon and 270% for the hand spouts just by changing the liquid hose position from the inner to the outermost spout position. Similar results were obtained for D10 and D90. In particular, the variation of liquid hose position from conventional to extreme out of spout position (XP) allowed to vary the spray quality generated from very fine (VF) to coarse (C)/very coarse (VC), giving farmers a wide range of options during the spray application. The droplet driftability, measured by the V100 parameter, decreased with the increase of the liquid hose distance from its original position. This suggests that when the liquid hose is in the extreme outer position, the pneumatic nozzle behaves similarly to a hydraulic drift reducing nozzle.

The findings of this study could significantly help in reducing the spray drift in pneumatic spraying just by slightly modifying the spray nozzle, which could have important practical implications. For the first time it is possible to design a device that gives farmers and technicians the possibility to match the drift-reduction environmental requirements using pneumatic sprayers for 3D crops, balancing at the same time the treatment specifications for every spray application.

Author Contributions: Conceptualization, M.G., A.M.F., P.M. and P.B.; methodology, M.G., A.F.M., P.M. and P.B.; validation, M.G., A.F.M. and P.M.; formal analysis, M.G. and A.F.M.; investigation, M.G., A.F.M., P.M. and F.G; resources, P.B.; data curation, M.G. and A.F.M.; writing—original draft preparation, M.G. and A.F.M.; writing—review and editing, M.G., A.F.M., P.M., P.B and F.G.; visualization, M.G.; supervision, M.G., P.B. and F.G.; project administration, F.G.; funding acquisition, P.B.; All authors have read and agreed to the published version of the manuscript.

Funding: This research received no external funding.

Acknowledgments: The authors would like to thank CIMA S.p.a. (Pavia, IT) for providing the spouts used in the experimental work.

Conflicts of Interest: The authors declare no conflict of interest.

Abbreviations

AS	Air speed
AP	Alternative insertion position of the liquid hose in the spout
CP	Conventional insertion position of the liquid hose in the spout
D10	Diameter for which a volume fraction of 10% is made up of drops with diameters smaller than this value (expressed in μm)
D50	Diameter for which a volume fraction of 50% is made up of drops with diameters smaller than this value (expressed in μm)
D90	Diameter for which a volume fraction of 90% is made up of drops with diameters smaller than this value (expressed in μm)
HP	Liquid hose position
LFR	Liquid flow rate in the spraying circuit
OP	Insertion position of the liquid hose out of the spout
RSF	Relative SPAN factor, a measure of the droplet homogeneity in the spray population (dimensionless)
V100	Spray liquid fraction generated with droplets smaller than 100 μm (expressed in %)
VMD	Volumetric median diameter, equivalent to D50
XP	Insertion position of the liquid hose at the extreme distance out of the spout

References

1. International Organization of Vine and Wine. OIV Statistical Report on World Vitiviniculture. 2018. Available online: http://www.oiv.int/public/medias/6371/oiv-statistical-report-on-world-vitiviniculture-2018.pdf (accessed on 7 September 2020).
2. Goodell, P.B. Fifty years of integrated control concept: The role of landscape ecology in IPM in San Joaquin valley cotton. *Pest Manag. Sci.* **2009**, *65*, 1293–1297. [CrossRef] [PubMed]
3. Gil, E.; Escolà, A. Design of a decision support method to determine volume rate for vineyard spraying. *Appl. Eng. Agric.* **2009**, *25*, 145–152. [CrossRef]
4. EC. Directive 2009/128/EC of the European Parliament and the Council of 21 October 2009 Establishing a Framework for Community Action to Achieve the Sustainable use of Pesticides. *Off. J. Eur. Union L* **2009**, *309*, 71–86.
5. Cerruto, E.; Manetto, G.; Longo, D.; Failla, S.; Papa, R. A model to estimate the spray deposit by simulated water sensitive papers. *Crop. Prot.* **2019**, *124*, 104861. [CrossRef]
6. Cross, J.V.; Walklate, P.J.; Murray, R.A.; Richardson, G.M. Spray deposits and losses in different sized apple trees from an axial fan orchard sprayer: 1. Effects of spray liquid flow rate. *Crop. Prot.* **2001**, *20*, 13–30. [CrossRef]
7. Cross, J.V.; Walklate, P.J.; Murray, R.A.; Richardson, G.M. Spray deposits and losses in different sized apple trees from an axial fan orchard sprayer: 2. Effects of spray quality. *Crop. Prot.* **2001**, *20*, 333–343. [CrossRef]
8. Cross, J.V.; Walklate, P.; Murray, R.; Richardson, G. Spray deposits and losses in different sized apple trees from an axial fan orchard sprayer: 3. Effects of air volumetric flow rate. *Crop. Prot.* **2003**, *22*, 381–394. [CrossRef]
9. Failla, S.; Romano, E.; Longo, D.; Bisaglia, C.; Schillaci, G. Effect of different axial fans configurations on airflow rate. In *Innovative Biosystems Engineering for Sustainable Agriculture, Forestry and Food Production, Proceedings of the MID-TERM AIIA, Matera, Italy, 12–13 September*; Coppola, A., Di Renzo, G., Altieri, G., D'Antonio, P., Eds.; Springer: Cham, Switzerland, 2020; Volume 67, pp. 691–699. [CrossRef]
10. Hołownicki, R.; Doruchowski, G.; Świechowski, W.; Godyń, A.; Konopacki, P.J. Variable air assistance system for orchard sprayers; concept, design and preliminary testing. *Biosyst. Eng.* **2017**, *163*, 134–149. [CrossRef]
11. Longo, D.; Manetto, G.; Papa, R.; Cerruto, E. Design and construction of a low-cost test bench for testing agricultural spray nozzles. *Appl. Sci.* **2020**, *10*, 5221. [CrossRef]
12. Nuyttens, D.; Baetens, K.; de Schampheleire, M.; Sonck, B. Effect of nozzle type, size and pressure on spray droplet characteristics. *Biosyst. Eng.* **2007**, *97*, 333–345. [CrossRef]
13. Salcedo, R.; Vallet, A.; Granell, R.; Garcerá, C.; Moltó, E.; Chueca, P. Eulerian–Lagrangian model of the behaviour of droplets produced by an air-assisted sprayer in a citrus orchard. *Biosyst. Eng.* **2017**, *154*, 76–91. [CrossRef]
14. Świechowski, W.; Doruchowski, G.; Holownicki, R.; Godyn, A. Penetration of air within the apple tree canopy as affected by the air jet characteristics and travel velocity of the sprayer. *Electron. J. Pol. Agric. Univ.* **2004**, *7*, #03.
15. Celen, I.H.; Arin, S.; Durgut, M.R. The effect of the air blast sprayer speed on the chemical distribution in vineyard. *Pak. J. Biol. Sci.* **2008**, *11*, 1472–1476. [CrossRef]
16. Cerruto, E. Influence of airflow rate and forward speed on the spray deposit in vineyards. *J. Agric. Eng.* **2007**, *1*, 7–14.
17. Duga, A.T.; Ruysen, K.; Dekeyser, D.; Nuyttens, D.; Bylemans, D.; Nicolaï, B.M.; Verboven, P. Spray deposition profiles in pome fruit trees: Effects of sprayer design, training system and tree canopy characteristics. *Crop. Prot.* **2015**, *67*, 200–213. [CrossRef]
18. ISO. *ISO22866:2005: Equipment for Crop Protection-Methods for Field Measurements of Spray Drift*; International Organization for Standardization: Geneva, Switzerland, 2005; pp. 1–17.
19. Arvidsson, T.; Bergström, L.; Kreuger, J. Spray drift as influenced by meteorological and technical factors. *Pest Manag. Sci.* **2011**, *67*, 586–598. [CrossRef]
20. Garcerá, C.; Moltó, E.; Chueca, P. Spray pesticide applications in Mediterranean citrus orchards: Canopy deposition and off-target losses. *Sci. Total Environ.* **2017**, *599*, 1344–1362. [CrossRef]

21. Gil, E.; Llorens, J.; Gallart, M.; Gil-Ribes, J.A.; Miranda-Fuentes, A. First attempts to obtain a reference drift curve for traditional olive grove's plantations following ISO 22866. *Sci. Total Environ.* **2018**, *627*, 349–360. [CrossRef] [PubMed]
22. Gil, E.; Llorens, J.; Llop, J.; Fàbregas, X.; Gallart, M. Use of a terrestrial LIDAR sensor for drift detection in vineyard spraying. *Sensors* **2013**, *13*, 516–534. [CrossRef]
23. Grella, M.; Gallart, M.; Marucco, P.; Balsari, P.; Gil, E. Ground deposition and airborne spray drift assessment in vineyard and orchard: The influence of environmental variables and sprayer settings. *Sustainability* **2017**, *9*, 728. [CrossRef]
24. Grella, M.; Marucco, P.; Balafoutis, A.T.; Balsari, P. Spray drift generated in vineyard during under-row weed control and suckering: Evaluation of direct and indirect drift-reducing techniques. *Sustainability* **2020**, *12*, 5068. [CrossRef]
25. Grella, M.; Marucco, P.; Manzone, M.; Gallart, M.; Balsari, P. Effect of sprayer settings on spray drift during pesticide application in poplar plantations (*Populus* spp.). *Sci. Total Environ.* **2017**, *578*, 427–439. [CrossRef]
26. Praat, J.P.; Maber, J.F.; Manktelow, D.W.L. The effect of canopy development and sprayer position on spray drift from a pipfruit orchard. *N. Z. Plant Prot.* **2000**, *53*, 241–247. [CrossRef]
27. Salyani, M.; Farooq, M.; Sweeb, R.D. Spray deposition and mass balance in citrus orchard applications. *Trans. ASABE* **2007**, *50*, 1963–1969. [CrossRef]
28. Felsot, A.S.; Unsworth, J.B.; Linders, J.B.H.J.; Roberts, G. Agrochemical spray drift; assessment and mitigation—A review. *J. Environ. Sci. Health B* **2011**, *46*, 1–23. [CrossRef]
29. Balsari, P.; Marucco, P.; Tamagnone, M. A system to assess the mass balance of spray applied to tree crops. *Trans. ASAE* **2005**, *48*, 1689–1694. [CrossRef]
30. Derksen, R.C.; Zhu, H.; Fox, R.D.; Brazee, R.D.; Krause, C.R. Coverage and drift produced by air induction and conventional hydraulic nozzles used for orchard applications. *Trans. ASABE* **2007**, *50*, 1493–1501. [CrossRef]
31. Bonds, J.A.S.; Leggett, M. A Literature Review of Downwind Drift from Airblast Sprayers: Development of Standard Methodologies and a Drift Database. *Trans. ASABE* **2015**, *58*, 1471–1477.
32. Fox, R.D.; Derksen, R.C.; Zhu, H.; Brazee, R.D.; Svensson, S.A. A history of air-blast sprayer development and future prospects. *Trans. ASABE* **2008**, *51*, 405–410. [CrossRef]
33. Gil, Y.; Sinfort, C. Emission of pesticides to the air during sprayer application: A bibliographic review. *Atmos. Environ.* **2005**, *39*, 5183–5193. [CrossRef]
34. Hewitt, A.J. Droplet size and agricultural spraying, part I: Atomization, spray transport, deposition, drift, and droplet size measurement techniques. *At. Sprays* **1997**, *7*, 235–244. [CrossRef]
35. Lešnik, M.; Pintar, C.; Lobnik, A.; Kolar, M. Comparison of the effectiveness of standard and drift-reducing nozzles for control of some pests of apple. *Crop. Prot.* **2005**, *24*, 93–100. [CrossRef]
36. Miranda-Fuentes, A.; Rodríguez-Lizana, A.; Cuenca, A.; González-Sánchez, J.; Blanco-Roldán, G.L.; Gil-Ribes, J.A. Improving plant protection product applications in traditional and intensive olive orchards through the development of new prototype air-assisted sprayers. *Crop. Prot.* **2017**, *94*, 44–58. [CrossRef]
37. Rautmann, D.; Streloke, M.; Winkler, R. New basic drift values in the authorization procedure for plant protection products. *Mitt. Biol. Bundesanst Forstwirtsch* **2001**, *383*, 133–141.
38. TOPPS-Prowadis Project. Best Management Practices to Reduce Spray Drift. 2014. Available online: http://www.topps-life.org/ (accessed on 7 September 2020).
39. Melese, E.A.; Debaer, C.; Rutten, N.; Vercammen, J.; Delele, M.A.; Ramon, H.; Nicolaï, B.M.; Verboven, P. A new integrated CFD modelling approach towards air-assisted orchard spraying. Part I. Model development and effect of wind speed and direction on sprayer airflow. *Comput. Electron. Agric.* **2010**, *71*, 128–136.
40. Herbst, A. A method to determine spray drift potential from nozzles and its link to buffer zone restrictions. In Proceedings of the ASAE Annual International Meeting, Sacramento, CA, USA, 28 July–1 August 2001. Paper number: 1-1047.
41. Doruchowski, G.; Swiechowski, W.; Holownicki, R.; Godyn, A. Environmentally-Dependent Application System (EDAS) for safer spray application in fruit growing. *J. Hortic. Sci. Biotechnol.* **2009**, *86*, 107–112. [CrossRef]
42. Hofman, V.; Solseng, E. *Reducing Spray Drift*; North Dakota State University NDSU Extension Service AE-1210; North Dakota State University NDSU Extension Service: Fargo, ND, USA, 2017.

43. Van de Zande, J.C.; Holterman, H.J.; Wenneker, M. Nozzle classification for drift reduction in orchard spraying: Identification of drift reduction class threshold nozzles. *Agric. Eng. Int. CIGR Ejournal* **2008**, *10*, 1–12.
44. Nuyttens, D.; de Schampheleire, M.; Verboven, P.; Brusselman, E.; Dekeyser, D. Droplet size and velocity characteristics of agricultural sprays. *Trans. ASABE* **2009**, *52*, 1471–1480. [CrossRef]
45. Garcerá, C.; Román, C.; Moltó, E.; Abad, R.; Insa, J.A.; Torrent, X.; Planas, S.; Chueca, P. Comparison between standard and drift reducing nozzles for pesticide application in citrus: Part II. Effects on canopy spray distribution, control efficacy of *Aonidiella aurantii* (Maskell), beneficial parasitoids and pesticide residues on fruit. *Crop. Prot.* **2017**, *94*, 83–96. [CrossRef]
46. Cawood, P.N.; Robinson, T.H.; Whittaker, S. An investigation of alternative application techniques for the control of blackgrass. In Proceedings of the Brighton Crop Protection Conference-Weeds, Brighton, UK, 20–23 November 1995; pp. 521–528.
47. Doruchowski, G.; Swiechowski, W.; Masny, S.; Maciesiak, A.; Tartanus, M.; Bryk, H.; Hołownicki, R. Low-drift nozzles vs. standard nozzles for pesticide application in the biological efficacy trials of pesticides in apple pest and disease control. *Sci. Total Environ.* **2017**, *575*, 1239–1246. [CrossRef] [PubMed]
48. Balsari, P.; Gil, E.; Marucco, P.; van de Zande, J.C.; Nuyttens, D.; Herbst, A.; Gallart, M. Field-crop-sprayer potential drift measured using test bench: Effects of boom height and nozzle type. *Biosyst. Eng.* **2017**, *154*, 3–13. [CrossRef]
49. Codis, S.; Verges, A.; Auvergne, C.; Bonicel, J.F.; Diouloufet, G.; Cavalier, R.; Douzals, J.P.; Magnier, J.; Montegano, P.; Ribeyrolles, X.; et al. Optimization of early growth stage treatments of the vine: Experimentations on the artificial vine EvaSprayViti. Proceedings of SuproFruit, 13th Workshop on Spray Application Techniques in Fruit Growing, Lindau/Lake Constance, Germany, 15–18 July 2015; pp. 47–48.
50. Porras Soriano, A.; Porras Soriano, M.L.; Porras Piedra, A.; Soriano Martìn, M.L. Comparison of the pesticide coverage achieved in a trellised vineyard by a prototype tunnel sprayer, a hydraulic sprayer, an air assisted sprayer and a pneumatic sprayer. *Span. J. Agric. Res.* **2005**, *3*, 175–181. [CrossRef]
51. Marucco, P.; Balsari, P.; Grella, M.; Pugliese, M.; Eberle, D.; Gil, E.; Llop, J.; Fountas, S.; Mylonas, N.; Tsitsigiannis, D.; et al. OPTIMA EU project: Main goal and first results of inventory of current spray practices in vineyards and orchards. In Proceedings of the SuproFruit, 15th Workshop on Spray Application and Precision Technology in Fruit Growing, East Malling, UK, 16–18 July 2019; pp. 99–100.
52. Grella, M.; Marucco, P.; Balsari, P. Evaluation of potential spray drift generated by different aypes of Airblast sprayers using an "ad hoc" test bench device. In *Innovative Biosystems Engineering for Sustainable Agriculture, Forestry and Food Production, Proceedings of the MID-TERM AIIA, Matera, Italy, 12–13 September*; Coppola, A., Di Renzo, G., Altieri, G., D'Antonio, P., Eds.; Springer: Cham, Switzerland, 2020; Volume 67, pp. 431–440. [CrossRef]
53. Balsari, P.; Grella, M.; Marucco, P.; Matta, F.; Miranda-Fuentes, A. Assessing the influence of air speed and liquid flow rate on the droplet size and homogeneity in pneumatic spraying. *Pest Manag. Sci.* **2019**, *75*, 366–379. [CrossRef]
54. Balsari, P.; Tamagnone, M.; Marucco, P.; Matta, F. How air liquid parameters affect droplet size: Experiences with different pneumatic nozzles. *Asp. Appl. Biol.* **2016**, *132*, 265–271.
55. Miranda-Fuentes, A.; Marucco, P.; Gonzalez-Sanchez, E.J.; Gil, E.; Grella, M.; Balsari, P. Developing strategies to reduce spray drift in pneumatic spraying vineyards: Assessment of the parameters affecting droplet size in pneumatic spraying. *Sci. Total Environ.* **2018**, *616–617*, 805–815. [CrossRef]
56. Byass, J.B.; Lake, J.R. Spray drift from a tractor-powered field sprayer. *Pestic. Sci.* **1977**, *8*, 117–126. [CrossRef]
57. Grover, R.; Kerr, L.A.; Maybank, J.; Yoshida, K. Field measurements of droplets drift from ground sprayers. I. Sampling, analytical and data integration techniques. *Can. J. Plant Sci.* **1978**, *58*, 611–622. [CrossRef]
58. Grella, M.; Miranda-Fuentes, A.; Marucco, P.; Balsari, P. Field assessment of a newly-designed pneumatic spout to contain spray drift in vineyards: Evaluation of canopy distribution and off-target losses. *Pest. Manag. Sci.* **2020**. [CrossRef] [PubMed]
59. Matthews, G.A.; Hislop, E.C. *Application Technology for Crop Protection*, 1st ed.; C.A.B. International: Wallingford, UK, 1993; p. 359.
60. Wenneker, M.; van de Zande, J.C.; Stallinga, H.; Michielsen, J.M.P.G.; van Velde, P.; Nieuwenhuitzen, A.T. Emission reduction in orchards by improved spray deposition and increased spray drift reduction of multiple row sprayers. *Asp. Appl. Biol.* **2014**, *122*, 195–202.

61. R Core Team. *R: A Language and Environment for Statistical Computing*; R Foundation for Statistical Computing: Vienna, Austria, 2019; Available online: https://www.R-project.org (accessed on 7 September 2020).
62. IBM Corp. *IBM SPSS Statistics for Windows*; Version 25.0; IBM Corp: Armonk, NY, USA, 2017.
63. American Society of Agricultural Engineers. *ASABE S572.1 Spray Nozzle Classification by Droplet Spectra*; American Society of Agricultural Engineers: St. Joseph, MI, USA, 2009; p. 4.
64. Southcombe, E.S.E.; Miller, P.C.H.; Ganzelmeier, H.; van de Zande, J.C.; Miralles, A.; Hewitt, A.J. The international (BCPC) spray classification system including a drift potential factor. In Proceedings of the Brighton Crop Protection Conference-Weeds, Brighton, UK, 17–20 November 1997; pp. 371–380.

Article

Spatial Distribution of Spray from a Solid Set Canopy Delivery System in a High-Density Apple Orchard Retrofitted with Modified Emitters

Rakesh Ranjan [1], Rajeev Sinha [1], Lav R. Khot [1,*], Gwen-Alyn Hoheisel [1], Matthew Grieshop [2] and Mark Ledebuhr [3]

1 Center for Precision and Automated Agricultural Systems, Biological Systems Engineering, Washington State University, Prosser, WA 99350, USA; rakesh.ranjan@wsu.edu (R.R.); rajeev.sinha@wsu.edu (R.S.); ghoheisel@wsu.edu (G.-A.H.)
2 Department of Entomology, Michigan State University, East Lancing, MI 48824, USA; grieshop@msu.edu
3 Application Insight, LLC, Lansing, MI 48906, USA; mark@applicationinsightllc.com
* Correspondence: lav.khot@wsu.edu; Tel.: +1-509-786-9302

Abstract: Solid Set Canopy Delivery Systems (SSCDS) are fixed agrochemical delivery systems composed of a network of micro-sprayers/nozzles distributed in perennial crop canopies. A previous SSCDS design composed of a 3-tier configuration using hollow cone sprayer nozzles has been shown to provide excellent coverage and deposition in high-density apple orchards. However, the hollow cone nozzles substantially increases the initial system installation costs. This study evaluated the effect of irrigation micro-emitters replacement on spray deposition, coverage and off-target drift. A micro-emitter used in greenhouse irrigation systems was duly modified to enhance its applicability with SSCDS. After laboratory assessment and optimization of the micro-emitters, a replicated field study was conducted to compare 3-tier SSCDS configured with either of modified irrigation micro-emitters or traditional hollow cone nozzles. Canopy deposition and off target drift were evaluated using a 500 ppm fluorescent tracer solution sprayed by the field installed systems and captured on mylar collectors. Spray coverage was evaluated using water sensitive papers. The overall canopy deposition and coverage for treatment configured with modified irrigation micro-emitters (955.5 ± 153.9 [mean ± standard error of mean] ng cm^{-2} and 22.7 ± 2.6%, respectively) were numerically higher than the hollow cone nozzles (746.2 ± 104.7 ng cm^{-2} and 19.0 ± 2.8%, respectively). Moreover, modified irrigation micro-emitter SSCDS had improved spray uniformity in the canopy foliage and on either side of leaf surfaces compared to a hollow cone nozzle. Ground and aerial spray losses, quantified as deposition, were numerically lower for the modified irrigation micro-emitter (121.8 ± 43.4 ng cm^{-2} and 0.7 ± 0.1 ng cm^{-2}, respectively) compared to the traditional hollow cone nozzle (447.4 ± 190.9 ng cm^{-2} and 3.2 ± 0.4 ng cm^{-2}, respectively). Overall, the modified irrigation micro-emitter provided similar or superior performance to the traditional hollow cone nozzle with an estimated 12 times reduction in system installation cost.

Keywords: fixed spray delivery; SSCDS; spray drift; deposition; coverage

Citation: Ranjan, R.; Sinha, R.; Khot, L.R.; Hoheisel, G.-A.; Grieshop, M.; Ledebuhr, M. Spatial Distribution of Spray from a Solid Set Canopy Delivery System in a High-Density Apple Orchard Retrofitted with Modified Emitters. *Appl. Sci.* **2021**, *11*, 709. https://doi.org/10.3390/app11020709

Received: 17 December 2020
Accepted: 7 January 2021
Published: 13 January 2021

1. Introduction

The United States (US) is the third largest producer of the apple (*Malus domestica*) in the world after China and European Union [1]. Total fresh market apple production in the US in the year 2018 was 3.4 million tons with a total worth 3 billion dollars—out of which 73% was produced in Washington State (WA) [2]. Commercial apple production requires numerous applications of agrochemicals including insecticides, fungicides, foliar nutrients, and plant growth regulators with the most common application equipment consisting of air-blast sprayers [3,4]. However, this technology has a high tendency to produce off-target spray drift, defined as the movement of sprayed droplets through the air away from the intended target [5]. Off target spray drift has been reported as a major contributor of environmental contamination and is among the top ten contributors causing human health risk around the world [3,6–8].

Increasing market demand, restricted labor availability, and mechanization advances have led to substantial modification to orchard systems with widespread transition to tall spindle, v-trellis, and bi-axis architectures [9–12]. Such architectural changes of moving from spherical to compact linear architectures have further intensified spray drift with traditional sprayers with large air volumes [12,13]. Growers have adopted several modified forms of air-assisted sprayers (e.g., vertical tower sprayer, tunnel sprayers, and electrostatic sprayers) which have demonstrated encouraging results in drift reduction. However, equipment size, maneuvering difficulties, high operational cost, and inconsistent performances based on canopy size are some of the reported difficulties associated with these technologies [3,14,15]. Tractor-based sprayers also contribute to soil compaction and crop loss due to physical impact between fruits and equipment [16,17]. Since heavy air-blast sprayers cannot be operated on saturated soils, critical agrochemical application timings can be missed, leading to crop loss [18]. Thus, there is a need and growing interest in the development of efficient spraying techniques designed specifically for modern orchard architectures. Recently, fixed spray application systems deemed Solid Set Canopy Delivery Systems (SSCDS) have been suggested as an alternative to tractor based sprayers for high density orchards and vineyards [12].

SSCDS have been evaluated for use in high-density apple orchards, vineyards, blueberries and other tree fruit systems with most of the work focusing on system development and measurement of deposition and coverage [12,18–23]. SSCDS pest management efficacy has been demonstrated for high-density apple orchards in Michigan and New York, USA [24–26]. A pneumatic spray delivery system was developed by Sinha et al. [21] to overcome the issue of non-uniformity in spraying associated with a hydraulic spray delivery approach. Efforts have also been made to automate the operational stages of a SSCDS for large-scale emplacements and commercial adaptation. Ranjan et al. [27] developed an electronic control system and a spray control unit for wireless and remote actuation of the SSCDS variant under study.

One of the key design constraints for SSCDS is that they rely on a large number of nozzles/micro-emitters (3000–10,400 per ha) and their placement within the canopy considerably affect the spray deposition and coverage [18,20,28]. For example, while a shower down configuration with a single nozzle atop each tree was reported as the simplest and most economical configuration, it provided reduced spray deposition in lower canopy regions and underside of leaves [22]. Another SSCDS configuration with hollow cone nozzles installed in a 3-tier (6 nozzles per tree) provided higher levels and more consistent spray deposition and coverage in a high-density apple orchard [28], but can cost prohibitive at approximately \$208,000 ha^{-1} (10,400 nozzle-assembly ha^{-1} at average \$20 nozzle-assembly^{-1}). The low-cost micro-emitters used in greenhouse irrigation may be an encouraging alternative to these nozzles. However, the spray attributes of such micro-emitters are not favorable for pesticide application in their current design. Therefore, this study evaluated the performance of a SSCDS with irrigation micro-emitters modified to mimic the spray attributes of a hollow cone nozzle. The specific objectives were to:

1. Design and evaluate a low-cost irrigation micro-emitter that mimics performance of a higher cost hollow cone nozzle.
2. Determine and compare the deposition, coverage, and off target drift performance of 3-tier SSCDS configuration that utilizes either modified irrigation micro-emitters or traditional hollow cone nozzles.

2. Materials and Methods

2.1. Micro-Emitter Modification

An impaction-style micro-emitter used in both greenhouse irrigation systems (model: Modular 7000, Jain Irrigation Inc., Fresno, CA, USA) and in previous SSCDS proof of concept experiments [22] was selected for modification (Figure 1a). Such micro-emitters consist of a static impaction plate which atomizes the columnar spray jet in a radial pattern with a large cone angle (150°) and wetted diameter (2100 mm), and marginal vertical throw (320 mm) (Figure 1b). The factory impaction plate has a toothed design (Figure 1a) that tends to coarsen the spray and direct it into non-uniform radial "rays" of spray. This creates a wide statistical span in the droplet spectra and irregular deposition. While this is desirable for irrigation to mitigate evaporation, finer droplets with a narrower span are desired for canopy applications.

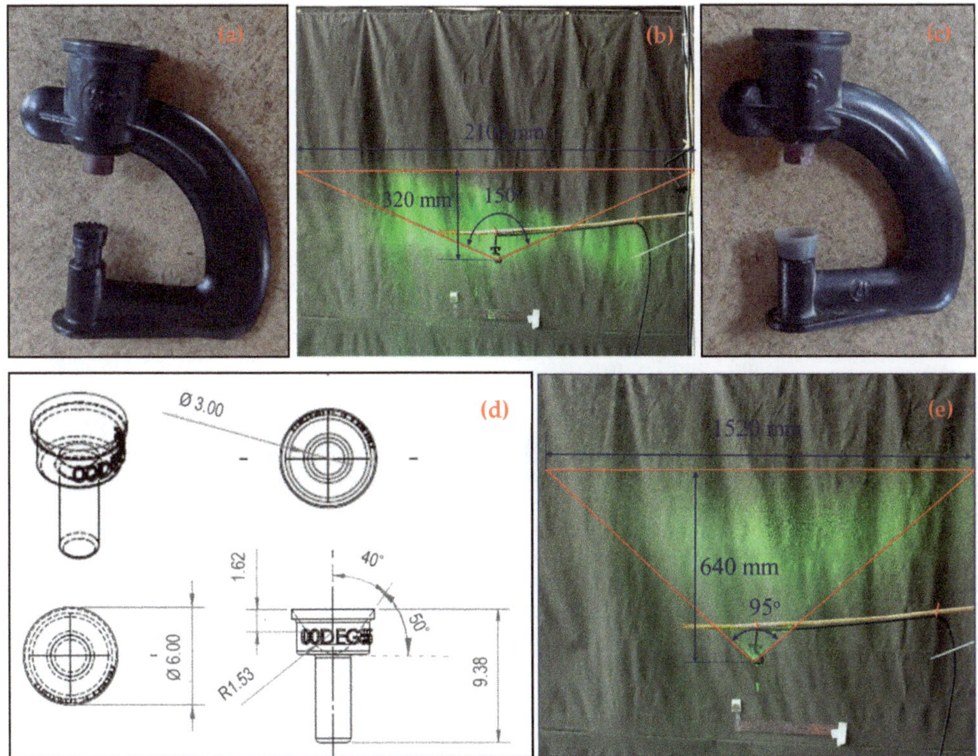

Figure 1. An off-the-shelf (**a**) micro-emitter with (**b**) a larger cone angle and wetted diameter, and a marginal vertical throw customized to (**c**) modified irrigation micro-emitter with (**d**) impactor plate concavity of 50° to acquire (**e**) a smaller cone angle and wetted diameter, and a higher vertical throw (not drawn to scale; all linear dimensions are in mm).

Through preliminary lab trials, it was hypothesized that increasing the static impactor plate concavity could reduce the cone angle of the spray, subsequently reducing the wetted diameter. Moreover, on vertically inverted placement of micro-emitter as depicted in Figure 1b, pertinent customization could increase the vertical throw of micro-emitters. Such modifications were critical to restrict the spray swath within the canopy, i.e., reduce off-target spray movement and increase in canopy deposition. Eliminating the teeth in the factory impaction plate and using a smooth-edged design also reduced the droplet spectrum, potentially improving in-canopy coverage. Thus, the static impactor plate of selected irrigation micro-emitter was modified with smooth edges and concavity ranging from 20–60° in an increment of 5°. The modified impactor plate was fixed to the micro-emitter assembly and evaluated in the lab. A randomly selected modified and non-modified micro-emitter was operated at 310 kPa, and a portable projector curtain was stationed in the background for imaging. A measuring scale was attached to the background for dimension referencing. The Red-Green-Blue images of the spray flux were captured using a visible-infrared sensor (model: Duo Pro R, FLIR Systems, Inc., Wilsonville, OR, USA) from a distance of 2 m and was analyzed in ImageJ (open source) software to evaluate the cone angle, vertical throw, and wetted diameter. The trial results indicated that a static spreader with 50° concavity (Figure 1d) was optimal to achieve the desired spray pattern with enhanced vertical throw and reduced wetted diameter (Figure 1c,e). Thus, the micro-emitter with modified static spreader (hereafter termed as 'modified irrigation micro-emitter') was selected for field evaluation in a SSCDS configuration. Additionally, the droplets of the micro-emitters/nozzles were characterized using a droplet size analyzer (model: VisiSize 15, Oxford Lasers Ltd., Didcot, Oxon, UK). The spray flux was passed through the optical sensing zone of the analyzer. The analyzer was set to analyze 1000 droplets, and the volume mean diameter ($D_V0.5$) corresponding to the micro-emitter/nozzle were evaluated. The droplet spectrum was classified based on the Dv0.5 values as per ASABE S572.3 standard [29].

2.2. *Field Trials*

2.2.1. SSCDS Spray Application System

A pneumatic spray delivery based SSCDS consisting of an applicator and a canopy delivery system (Figure 2) was selected for field trials [21]. Pertinent details regarding the applicator sub-systems with on-board pump, air-compressor, and spray tank can be found in Sinha et al. [18]. The canopy delivery sub-system consists of spray lines (main and return; ϕ = 2.54 cm), reservoir, and nozzle/micro-emitter assembly. Spray lines (main and return) were mounted on the existing orchard trellis wires at 1.4 m and 0.6 m above ground level, respectively, using poly hose trellis wire clips (ϕ = 2.54 cm, model: A32H, Jain Irrigation Inc., Fresno, CA, USA). The spray lines were connected in a loop and had manual flow control valves installed at the end of the loop. The reservoirs were mounted on the return line at an interval of 1.8 m. Each reservoir consisted of an inlet port, a bleed valve, a liquid column, an outlet port, a float, a diaphragm check valve, a nozzle supply column, pair of nozzle feed line, and an auto drain valve (Figure 2). These micro-emitters/nozzles were connected with the nozzle feed line of the reservoir using PE tubing (ϕ = 0.6 cm). The details of the micro-emitters/nozzles used in the two treatments are provided in Table 1.

The pneumatic spray delivery system has 3 operational stages, namely charging, recovery, and spraying/cleaning. In the charging stage, the reservoirs are filled with the spray mix using a hydraulic pressure of around 100 kPa through the main line. During recovery, the excess spray mix from the spray lines are recovered back to the spray tank using compressed air at 100 kPa through the return line. At this point, only the reservoirs contained spray mix and the contained volume was equivalent to one third of the application rate (234 L ha^{-1}). A diaphragm check valve (cracking pressure = 207 kPa) in the reservoir restricted any flow of spray mix through emitters during charging and recovery. After recovery, the spray mix contained in the reservoirs was sprayed under a pneumatic pressure of about 310 kPa. Once the spraying is complete, the auto drain valve in the reservoir opens to drain the residual volume onto the soil and cleaning is achieved.

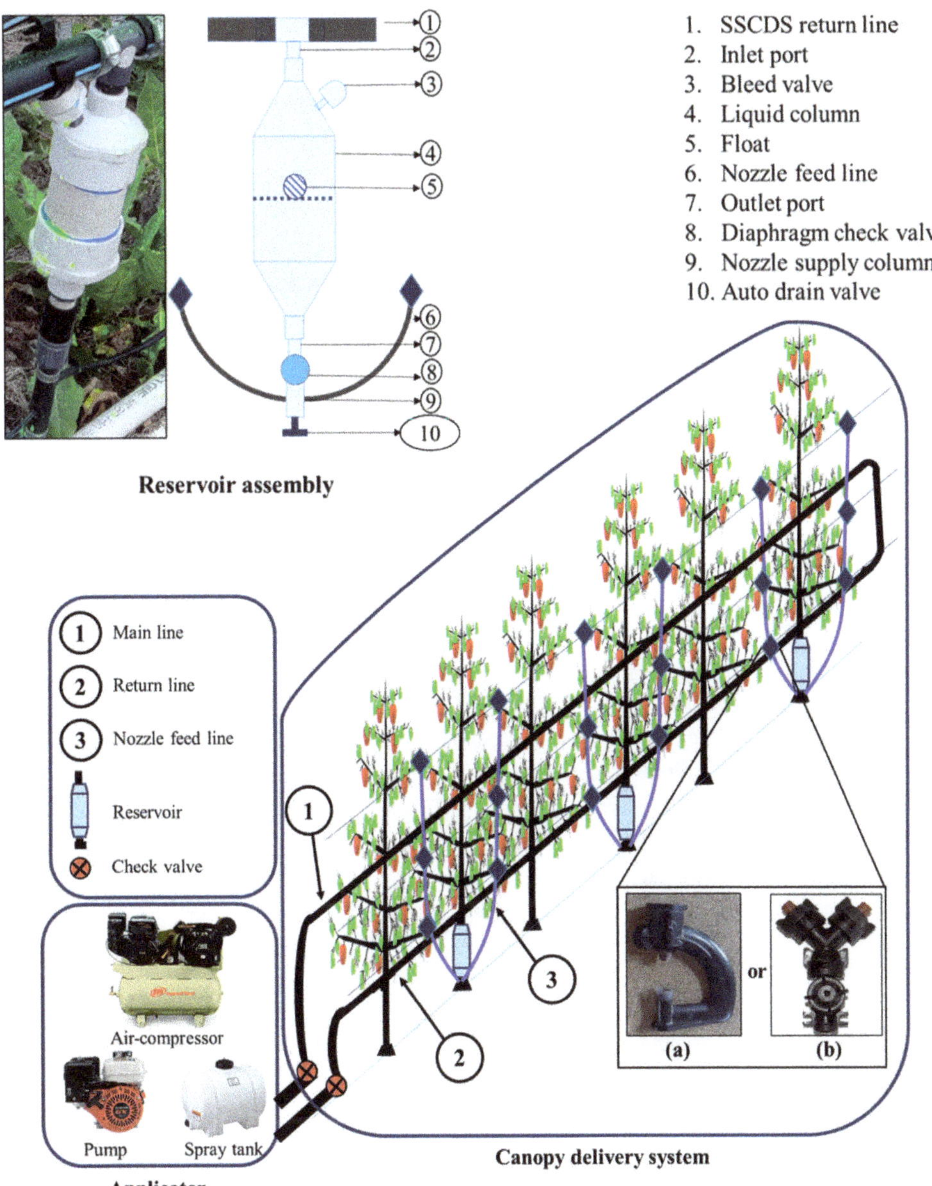

Figure 2. The schematics of tested pneumatic spray delivery based SSCDS with an applicator and canopy delivery system configured with (**a**) modified irrigation micro-emitter or (**b**) hollow cone nozzle in 3-tier arrangement (diamond shape with solid blue fill represents micro-emitter/nozzle).

Table 1. Specification of emitters tested in this study.

Emitter	Model	Manufacture	Spray Pattern	Flow Rate (L min^{-1})	Cost (USD $/Unit)
Modified micro-emitter	Modified	Jain Irrigation Inc.	Hollow cone	0.9	1.5
Hollow cone nozzle	TXVS12	TeeJet Technologies	Hollow cone	0.8	20.0

2.2.2. Treatment Details

Treatment T1, or 'irrigation micro-emitter treatment', was SSCDS configured with modified micro-emitters (Figure 2a) installed in a 3-tier arrangement (i.e., 3 micro-emitters per tree) (Table 1). The micro-emitters were installed between two trees on the existing orchard trellis wires at 1.0 m, 1.8 m and 2.6 m above ground level using the self-locking zip tie wire. Installation insured that the spray was directed upward into the canopy and provided spray coverage to one-third of the tree canopy. Sinha et al. [18] observed that directing spray upward into the canopy was critical to achieve spray coverage and deposition on abaxial leaf surfaces. The treatment T2 or 'hollow cone nozzle treatment' was also a pneumatic spray delivery based SSCDS, with emitter arrangement similar to T1. However, the micro-emitters were substituted with a pair of off-the-shelf hollow cone nozzles (TXVS12, TeeJet Technologies, Wheaton, IL, USA) connected to the spray line using a Y shaped quick-connect adapter (adapter: QJ90–2–NYR, nozzle body: QJ98590, TeeJet Technologies, Wheaton, IL, USA), with two mirrored spray outlets at 45° (Figure 2b). The quick-connect adapters were secured at each location (i.e., 1.0 m, 1.8 m and 2.6 m above ground level) to a PVC support pipe (ϕ = 1.3 cm) which was installed midway between two trees.

2.2.3. Study Site and Experimental Plot Layout

The spray trials were conducted in an apple orchard (cv. Cosmic Crisp) planted on M9-NIC29 rootstock in year 2013 and was trained in a tall spindle architecture. The research orchard was located in Roza Farm (46.29° N, 119.73° W) of Washington State University. The planting density of the experimental plot was 4284 tree ha^{-1} with an inter-row spacing of 3 m, plant to plant distance of 0.9 m, and mean tree height of 3 m.

A set of 11 apple trees planted between two wooden posts, positioned 10 m apart (hereafter termed as blocks), were designated for the system installation. Out of the 35 blocks in the experimental orchard, 6 were randomly selected for the spray trials and, in each of the blocks, a 10 m long pneumatic spray delivery based SSCDS was installed with modified micro-emitters or hollow cone nozzles in a 3-tier arrangement. Three blocks were treated with a modified irrigation micro-emitter (T1), while the other three blocks were treated with a hollow cone nozzle SSCDS (T2) (Figure 3). To ensure that the two treatments did not interact, the treatment specific spray trials were conducted on two different dates (i.e., T1: 22 July 2019 and T2: 24 July 2019). On a given day, three replicate trials were conducted within 45 min of time window.

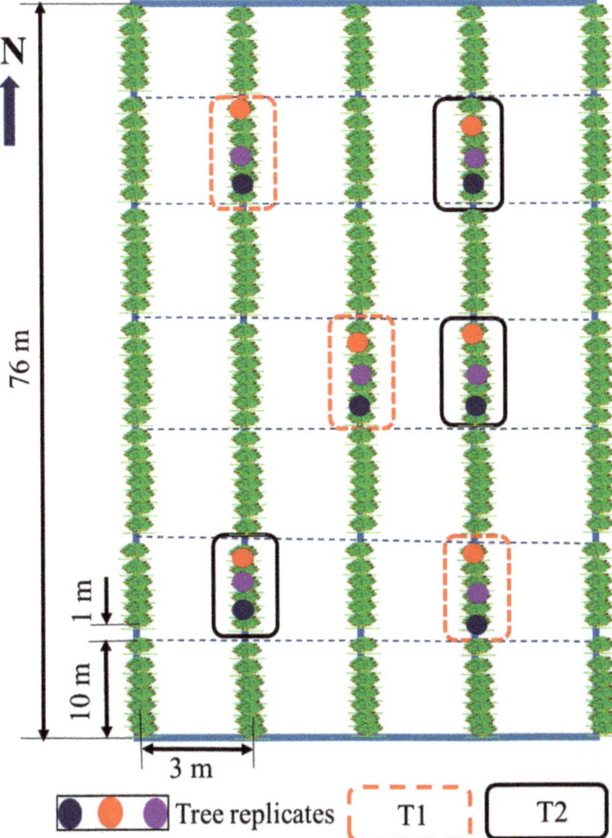

Figure 3. The schematics of experimental plot (not drawn to the scale). The dotted red boxes indicate the blocks treated with irrigation micro-emitter (T1), the solid blue boxes indicate the blocks with hollow cone nozzle treatment (T2), and the circular dots represents the sampled trees within a block.

2.2.4. Experimental Design

Spray Deposition and Coverage Evaluation

The spray trials quantified spray deposition and coverage following a randomized split-split plot design. Mylar cards (size: 5.1 × 5.1 cm, Stark Boards, CA, USA) and water sensitive papers (WSP) (size: 2.5 × 2.5 cm, Syngenta Crop Protection Inc., Greensboro, NC, USA) were used to quantify spray deposition and coverage, respectively (Figure 4). The spray deposition was enumerated by evaluating the amount of active ingredient deposited on the unit area of the mylar card (ng cm^{-2}). The spray coverage was defined as the percentage area of the WSPs stained by the spray mix. Three trees were randomly selected from the treatment blocks (Figure 3), and the sampling trees were divided into east and west canopy sides. The canopy was further divided in three zones (bottom: 0.6 to <1.4 m, mid: 1.4 to <2.2 m and top: 2.2 to 3.0 m) that resulted in six sampling zones per tree (top-east, top-west, mid-east, mid-west, bottom-east, and bottom-west) (Figure 4e). In each of the sampling zones, two leaves were randomly selected to install mylar card and WSP samplers. The samplers were installed in each of the canopy zones by clamping them onto the adaxial and abaxial leaf surfaces using customized alligator clips. The active surface of WSPs were oriented upward and downward at adaxial and abaxial leaf surfaces, respectively (Figure 4d). A total of 108 mylar cards and WSP samplers (3 blocks × 3 trees/block × 2 sides/tree × 3 zones/side × 2 leaf surface/zone × 1 sampler/leaf surface) were collected for each treatment under study.

Figure 4. The field installation of (**a**) modified irrigation micro-emitter and (**b**) off-the-shelf nozzle utilized for respective SSCDS treatments, and the deposition and coverage analysis of the tested treatments with help of the (**c**) mylar card and the (**d**) water sensitive paper samplers installed in (**e**) three canopy zones on east and west side (top-east, top-west, mid-east, mid-west, bottom-east, and bottom-west).

Off-Target Spray Losses

Off-target spray losses were assessed in line with the randomized plots of canopy evaluations. Run-off and drift deposited on the ground and losses in the air were evaluated based on the schematic depicted in Figure 5. Sub-tree run-off includes the spray deposited underneath the trees because of the spray droplets settled to the tree bottom under gravity, rebounded droplets from the canopy and run-off due to canopy saturation. The run-off was evaluated with mylar card samplers (size: 5.1 × 5.1 cm) installed on a wooded block (size: 10 × 10 cm) placed below the replicate trees. Similarly, the downwind mid-row ground drift losses were evaluated using mylar card samplers located at a distance of 1.5, 4.5 and 7.5 m from the block being sprayed (Figure 5). The aerial drift losses were assessed by evaluating the tracer deposition above the tree canopy downwind to the block being sprayed. A customized PVC mast was utilized to hold mylar card samplers at a height of 3.3, 3.6 and 3.9 m above ground level. Two masts, carrying three samplers, were positioned 3 and 6 m downwind to evaluate the aerial drift losses (Figure 5). In addition to mylar card samplers, WSP samplers (size: 2.5 × 2.5 cm) were also installed at each of the drift quantification location to cross verify any contamination of mylar card samplers while handling and analysis [21].

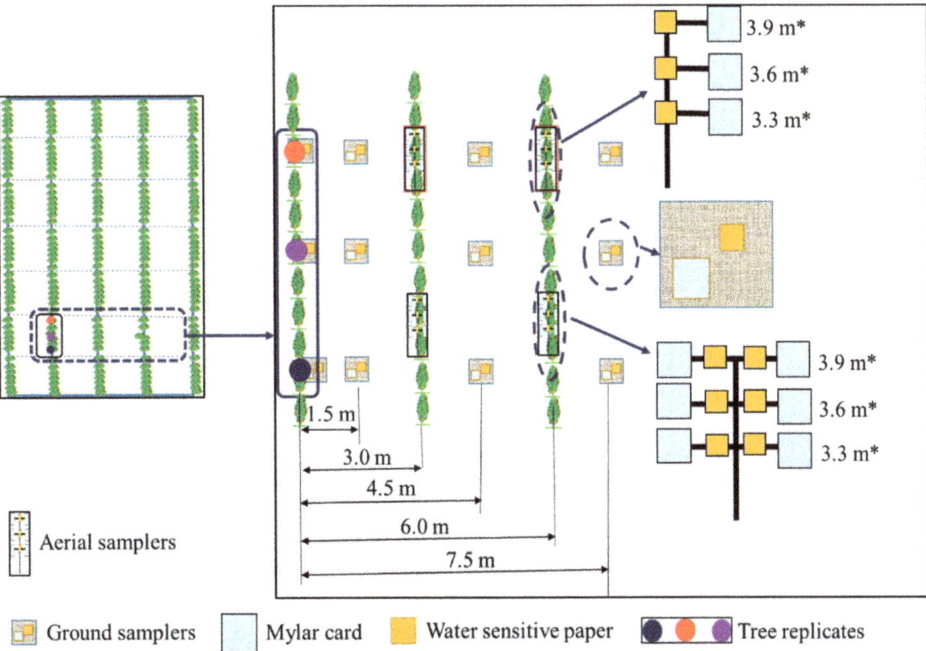

Figure 5. The off-target drift sampler layout (schematic not drawn on the scale) with ground samplers consists of a mylar card and water sensitive paper (WSP) kept on a wooden block beneath the tree and in the mid-row alley at a distance of 1.5 m, 4.5 m and 7.5 m away from the treated canopy. Three replicates of aerial drift samplers fixed on PVC mast at 3 m and 6 m downwind with mylar cards and WSPs fixed at a height of 3.3 m, 3.6 m and 3.9 m above the ground level (* distance measured above ground level).

2.3. Data Collection Protocol

A 500 ppm solution of Pyranine, a biodegradable fluorescent tracer (10G®, Keystone Inc., Chicago, IL, USA) was prepared with tap water. The spray mix was agitated thoroughly to create a homogeneous solution. Tank samples were collected pre- and post-spraying to monitor any change in tracer concentration during spraying and subsequent normalization of field samples [20].

The mylar cards and WSP samplers were installed in respective sampling zones prior to spray application as discussed in the Section 2.2.4. Since the pneumatic spray delivery based SSCDS used in this study was designed for 234 L ha^{-1}, the system was operated three times to obtain an application rate of 702 L ha^{-1}. The operating pressure during spraying stage was set at 310 kPa [21,22]. A pair of thin film pressure transducer (model: 1502B81EZ100psiG, PCB Piezotronics Inc., Depew, NY, USA) coupled with a data logger (model: CR1000, Campbell Scientific, Logan, UT, USA) was installed at 1.5 and 22 m away from the inlet port to log pressure data at 1 Hz. Additionally, an all-in-one weather station (model: ATMOS 41, METER Group, Inc., Pullman, WA, USA) coupled with a data logger (model: CR1000, Campbell Scientific, Logan, UT, USA) was installed at a height of 1 m above the canopy (ISO 22522, 2007) (Table 2) to monitor the in-field weather parameters. The weather parameters were logged at 0.2 Hz.

After spraying, the mylar card and WSPs samplers were allowed to dry for 15 min and collected and stored according to protocol described in Sinha et al. [20,22].

Table 2. Weather parameters recorded during the field data collection.

Treatment	Date	Trial	Weather Parameters (Mean ± Std. Dev.)			
			Wind Speed (m s^{-1})	Wind Direction (°) *	Air Temperature (°C)	Relative Humidity (%)
Irrigation micro-emitter	22 July 2019	1	1.0 ± 0.2	297.3 ± 13.6	18.7 ± 0.4	44.8 ± 1.5
		2	0.7 ± 0.2	219.0 ± 27.9	19.4 ± 0.1	48.4 ± 0.4
		3	0.6 ± 0.3	216.7 ± 37.0	21.0 ± 0.2	49.2 ± 1.0
Hollow cone nozzle	24 July 2019	1	1.2 ± 0.6	271.1 ± 20.8	14.1 ± 0.1	57.1 ± 0.6
		2	0.9 ± 0.4	254.4 ± 32.8	15.8 ± 0.1	57.3 ± 1.1
		3	1.4 ± 0.4	218.9 ± 18.9	17.2 ± 0.2	50.8 ± 0.4

* Reported with reference to true north and the tree rows were oriented north–south.

2.4. Data Analysis

The mylar cards and WSP samplers were analyzed using fluorometry analysis and image processing, respectively. The analysis was conducted in accordance with Sinha et al. [23] to estimate the tracer deposition per unit area (ng cm^{-2}) (hereafter termed as 'deposition') on the mylar card and spray coverage (%) (hereafter termed as 'coverage') on the WSP samplers.

The deposition and coverage data were analyzed in R studio (2017, version: 3.4.1) [30]. The datasets were cube root transformed for normalization. The transformed data were analyzed using a 2 × 3 × 2 factorial analysis of variance (ANOVA) with treatment (modified irrigation micro-emitter and hollow cone nozzle SSCDS), canopy zone (top, mid, and bottom), and leaf surface (adaxial and abaxial) as fixed factors. A Tukey Honest Significance Difference (HSD) post-hoc test was performed for multiple comparisons. The coefficient of variation (CV) in spray deposition along the leaf surface was evaluated to assess the spray uniformity for the tested treatments. Separate ANOVA models were run for the sub-tree run-off, mid-row ground, and aerial drift with deposition and coverage as the response variables. Pertaining to this, the treatments (modified irrigation micro-emitter and hollow cone nozzle SSCDS), downwind ground sampler distance (1.5 m, 4.5 m and 7.5 m), and aerial sampler height above the ground (3.3, 3.6 and 3.9 m) were used as a fixed factor. A confidence level of 95% was considered in all analyses.

3. Results

3.1. Spray Droplet Characterization

The customization of the micro-emitter resulted in reduced cone angle and wetted parameter and enhanced vertical throw for modified irrigation micro-emitter (concavity = 50°, cone angle = 95°, vertical throw = 640 mm and wetted diameter = 1520 mm) (Figure 1c,e) compared to the original micro-emitter (cone angle = 150°, vertical throw = 320 mm and wetted diameter = 2100 mm). Furthermore, modified micro-emitters produced a medium droplet size spectrum ($D_V0.5$ = 256.2 μm), while a hollow cone nozzle resulted in a fine droplet size spectrum ($D_V0.5$ = 130.9 μm) (Table 3). The pressure data indicate a marginal drop of 2 kPa between the main inlet and return outlet for both treatments. Such a small pressure drop eliminates the chances of variation in droplet characteristics due to reduction in pressure along the spray line.

Table 3. The volumetric droplet size distribution of the modified irrigation micro-emitter and hollow cone nozzle.

Emitter	Volume Percentile Diameter (μm) #			Category *
	$D_V0.1$	$D_V0.5$	$D_V0.9$	
Modified micro-emitter	134.6	256.2	416.8	Medium
Hollow cone nozzle	72.5	130.9	360.9	Fine

Reported volume percentile diameter at a pressure of 310 kPa, * Droplets have been categorized based on $D_V0.5$ as per ASABE S-572.3 standard [29].

3.2. Canopy Deposition

There were no significant differences in deposition for the main effects of SSCDS treatments ($F_{1,108}$ = 0.49, p = 0.48), canopy zones ($F_{2,72}$ = 0.23, p = 0.79) nor leaf surface ($F_{1,108}$ = 1.14, p = 0.29) (Table 4) from ANOVA. Likewise, no significant interaction effects were detected. Although not significant, the modified irrigation micro-emitter treatment provided numerically higher overall spray deposition (955.5 ± 153.9 ng cm^{-2}) (mean ± standard error of mean [SEM]) compared to the hollow cone nozzle SSCDS (746.2 ± 104.7 ng cm^{-2}) (Figure 6).

Table 4. ANOVA of cube root transformed canopy deposition data.

Variables	df	MS	F	p
Main plot				
Block	2	34.53		
Treatment	1	7.46	0.49	0.48
Error (a)	1	97.21		
Canopy zone	2	3.56	0.23	0.79
Leaf surface	1	17.34	1.14	0.29
Treatment × Canopy zone	2	23.07	1.52	0.22
Treatment × Leaf surface	1	1.51	0.1	0.75
Canopy zone × Leaf surface	2	20.12	1.32	0.268
Treatment × Canopy zone × Leaf surface	2	4.12	0.27	0.76
Error (b)	192	15.2		

3.2.1. Canopy Zone Level Deposition

ANOVA indicates non-significant differences in spray deposition among the canopy zones for both SSCDS treatments (Table 4). Moreover, there was no significant interaction effect between treatments and canopy zones. The bottom zone deposition for modified irrigation micro-emitter treatment i.e., T1 (1630.5 ± 401.1 ng cm^{-2}) was the highest followed by the bottom zone deposition for hollow cone nozzle treatment i.e., T2 (1022.3 ± 209.2 ng cm^{-2}) (Figure 6a). The least spray deposition was reported for the top canopy zone in treatment T1 (493.1 ± 117.6 ng cm^{-2}) and was significantly different than bottom zone deposition of corresponding treatment. These results indicate that similar deposition in different canopy zones may be achieved with modified irrigation micro-emitter SSCDS.

3.2.2. Leaf Surface Level Deposition

Spray deposition data collected at samplers installed on abaxial and adaxial surface of the leaves revealed that there was no significant difference in spray deposition on either surface of leaves regardless of the SSCDS treatments. Moreover, there was no significant interaction between treatment and leaf surface (Table 4). The highest spray deposition was reported for adaxial leaf surface treated with modified irrigation micro-emitter i.e., T1 (1112.3 ± 242.2 ng cm^{-2}) followed by hollow cone nozzle SSCDS i.e., T2 (914.5 ± 167.1 ng cm^{-2}) (Figure 6b). The abaxial deposition in T1 (798.7 ± 190.1 ng cm^{-2}) was also numerically higher than T2 (577.9 ± 123.4 ng cm^{-2}). Nevertheless, the differences were not significant with an HSD test. Furthermore, the CV in spray deposition on the leaf surfaces for T1 and T2 were 23.2% and 31.9%, respectively.

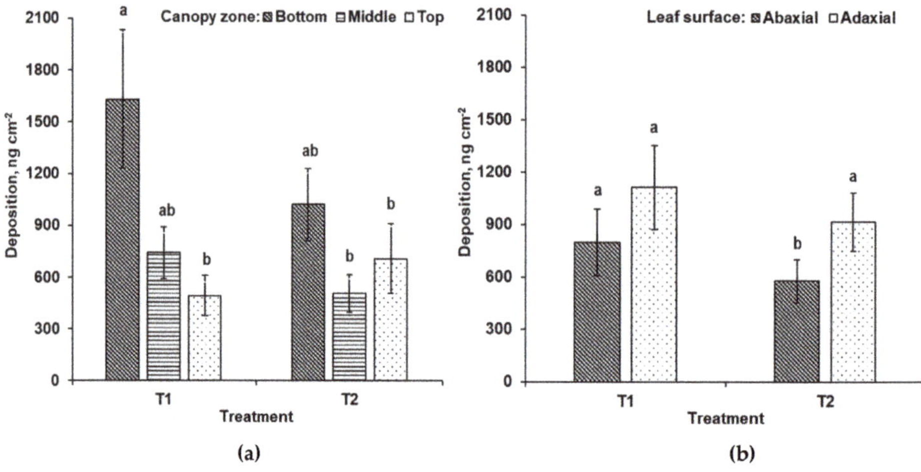

Figure 6. Mean spray deposition evaluated at (**a**) different canopy zones and (**b**) leaf surfaces for modified irrigation micro-emitter (T1) and hollow cone nozzle (T2) treatment. Different lowercase letters above individual bar plots indicate significance of mean differences in transformed data at 5% level and associated error bars indicate standard error; non-transformed deposition (ng cm^{-2}) values are presented.

3.3. Canopy Coverage

There was no significant difference in coverage corresponding to the SSCDS treatment ($F_{1,108} = 1.2, p = 0.27$) and canopy zones ($F_{2,72} = 0.36, p = 0.70$) (Table 5) as main effects. On the contrary, a significant coverage difference was reported between the leaf surfaces ($F_{1,108} = 8.91, p = 0.02$). Furthermore, there was no significant interaction between SSCDS treatments, canopy zones, and leaf surfaces. Overall, modified irrigation micro-emitter had numerically higher canopy coverage (22.7 ± 2.6%) compared to hollow cone nozzle SSCDS (19.0 ± 2.8%).

Table 5. ANOVA result of cube root transformed canopy coverage data.

Variables	df	MS	*F*	*p*
Main plot				
Block	2	0.04		
Treatment	1	2.00	1.20	0.27
Error(a)	1	9.85		
Canopy zone	2	0.60	0.36	0.70
Leaf surface	1	1.00	8.91	0.02
Treatment × Canopy zone	2	3.30	1.99	0.14
Treatment × Leaf surface	1	2.24	1.35	0.24
Canopy zone × Leaf surface	2	0.42	0.25	0.78
Treatment × Canopy zone × Leaf surface	2	0.68	0.41	0.66
Error(b)	192	1.66		

3.3.1. Canopy Zone Level Coverage

There was no significant difference in spray coverage among canopy zones regardless of the SSCDS treatments (Table 5). Moreover, no significant interaction was reported between canopy zone and SSCDS treatment. The bottom zone coverage of modified irrigation micro-emitter treatment i.e., T1 (34.6 ± 5.3%) was highest followed by bottom zone coverage of hollow cone nozzle treatment i.e., T2 (26.0 ± 5.7%) (Figure 7a). Moreover, top zone coverage of T1 (15.5 ± 3.9%) was significantly lower than bottom zone coverage. The least canopy coverage was reported for mid zone canopy coverage of treatment T2

(15.3 ± 3.9%). Furthermore, unlike T1, the differences in bottom and top zone coverage (15.8 ± 4.0%) for T2 were statistically non-significant.

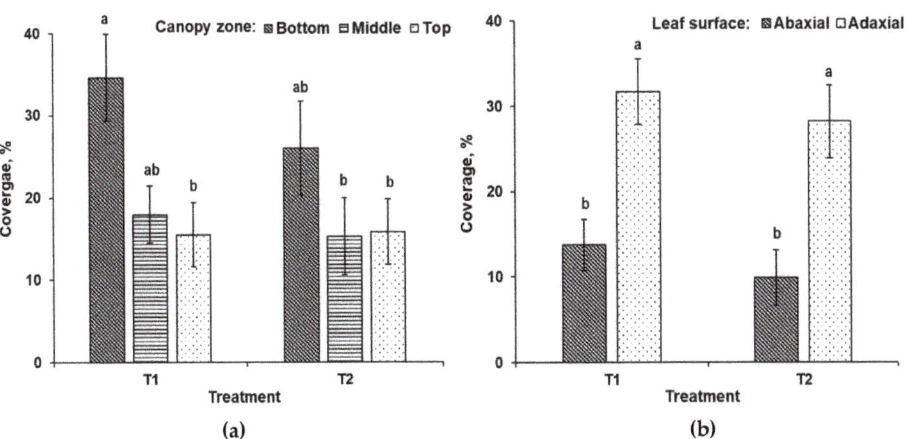

Figure 7. Mean spray coverage assessed at different (**a**) canopy zones and (**b**) leaf surfaces for modified irrigation micro-emitter (T1) and hollow cone nozzle (T2) treatments. The lowercase letters above individual bar plots indicate the significance of mean differences in transformed mean at 5% level and associated error bars indicate standard error; non-transformed coverage (%) values are presented.

3.3.2. Leaf Surface Level Coverage

There was a significant difference in abaxial and adaxial sample coverage for both SSCDS treatments (Figure 7b). However, interaction effect between treatment and surface was not significant (Table 5). The adaxial leaf surfaces of the canopies treated with modified irrigation micro-emitter (T1) received the highest spray coverage (31.7 ± 3.9%) followed by the adaxial leaf surfaces (28.2 ± 4.3%) of hollow cone nozzle SSCDS (T2) treated canopies. Moreover, abaxial coverage for T1 (13.7 ± 3.0%) and T2 (9.9 ± 3.2%) was significantly lower than corresponding adaxial coverage. Further analysis indicates that the CV in spray coverage among the leaf surfaces for T1 and T2 were 55.9% and 68.1%, respectively.

3.4. *Off-Target Drift Losses*

3.4.1. Ground Run-Off and Drift Losses

There was no significant difference in deposition for sub-tree run-off obtained from modified irrigation micro-emitter i.e., T1 (1720.6 ± 289.3 ng cm^{-2}) and hollow cone nozzle treatment i.e., T2 (1785.3 ± 435.6 ng cm^{-2}) (Table 6). Moreover, the percent of applied active ingredient lost underneath the tree for T1 (45.3%) was marginally lower than T2 (47.4%). The analysis of coverage samplers exhibited similar results. The treatment T1 and T2 had a coverage of 36.9 ± 5.7% and 35.3 ± 7.6%, respectively.

Mid-row ground drift deposition data collected at 1.5, 4.5 and 7.5 m downwind indicate that mean ground deposition for modified irrigation micro-emitter treatment, i.e., T1 (121.8 ± 43.4 ng cm^{-2}), was numerically lower than hollow cone nozzle SSCDS, i.e., T2 (447.4 ± 190.9 ng cm^{-2}) (Table 7). However, the difference among them were non-significant. Additionally, the percent of applied active ingredient lost to the ground drift for T1 (3.2 %) was considerably lower than T2 (20.8%). Furthermore, T1 had significantly lower mid-row ground deposition (364.4 ± 85.3 ng cm^{-2}) at 1.5 m downwind distance compared to T2 (1306.9 ± 465.3 ng cm^{-2}) (Figure 8a). The measured mid-row ground deposition at 4.5 and 7.5 m downwind for treatment T1 was also lower than T2; however, the difference was not significant. Similar trends were observed for the analysis of coverage samplers. The mean coverage corresponding to T1 (4.8 ± 1.7%) was lower than T2 (20.5 ± 6.2%); however, the difference was not significant (Table 7). Likewise, the coverage observed at

1.5 m downwind distance for treatment T1 (14.3 ± 3.3%) was significantly lower than T2 (61.0 ± 8.0%) (Figure 8b).

Table 6. Mean sub-tree run-off evaluated for tested treatments.

Off-Target Loss	Treatment	Deposition (ng cm^{-2}) * [Run-Off (%)] #	Coverage (%) *
Run-off	T1	1720.6 ± 289.3 ^{a} [45.3]	36.9 ± 5.7 ^{A}
	T2	1785.3 ± 435.6 ^{a} [47.4]	35.3 ± 7.6 ^{A}

* Transformed data were used for statistical analysis; presented data are non-transformed values in (mean ± SEM) format; different lowercase and uppercase letters represent the differences (significant or not significant) in transformed mean at α = 0.05; # The values in the square bracket represents the percent of applied active ingredient lost underneath the tree.

Table 7. Mean mid-row ground drift losses evaluated for tested treatments.

Off-Target Losses	Treatment	Deposition (ng cm^{-2}) * [Ground Drift (%)] #	Coverage (%) *
Mid-row ground drift	T1	121.8 ± 43.4 ^{a} [3.2]	4.8 ± 1.7 ^{A}
	T2	447.4 ± 190.9 ^{a} [20.8]	20.5 ± 6.2 ^{A}

* Transformed data was used for statistical analysis; presented data are non-transformed values in (mean ± SEM) format; different lowercase and uppercase letters represent the differences (significant or not significant) in transformed mean at α = 0.05; # The values in the square bracket represents the percent of applied active ingredient drifted on the ground. The lowercase and uppercase letters in superscript indicates the significant differences in transformed mean at α = 0.05.

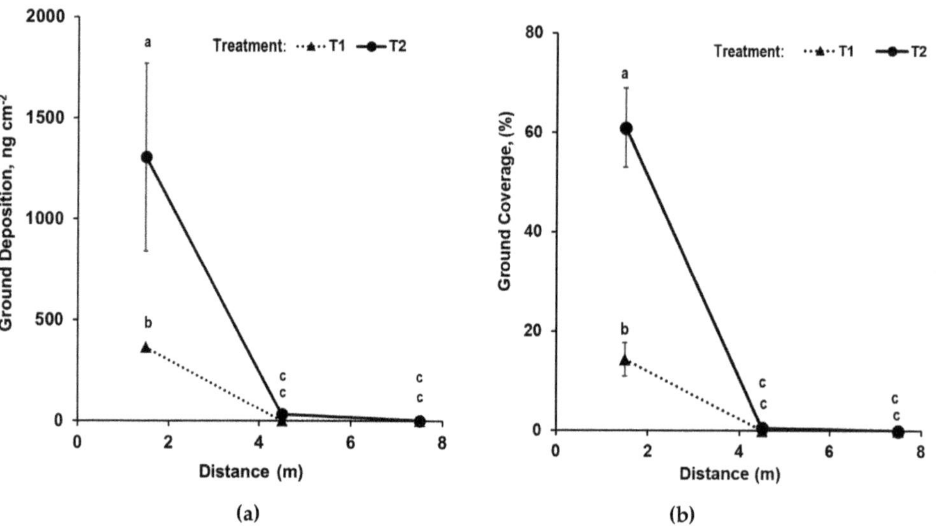

Figure 8. Mean mid-row ground (**a**) deposition and (**b**) coverage assessed at 1.5 m, 4.5 m and 7.5 m downwind for tested SSCDS treatments (i.e., T1 and T2). The lowercase letters above the line markers indicate the significant differences in transformed mean at α = 0.05 and associated error bars indicate standard error; presented values are non-transformed mid-row ground deposition (ng cm^{-2}) and coverage (%).

3.4.2. Aerial Drift Losses

Mean aerial deposition for modified irrigation micro-emitter i.e., T1 (0.7 ± 0.1 ng cm^{-2}) was significantly lower than hollow cone nozzle SSCDS treatment, i.e., T2 (3.2 ± 0.4 ng cm^{-2}) (Table 8). Additionally, the percent of applied active ingredient lost to the aerial drift was negligible for both treatments (0.02, and 0.08% for T1 and T2, respectively). Similar aerial deposition trends were observed at 3 and 6 m downwind (Figure 9a). Additionally, the deposition evaluated at various sampling heights (i.e., 3.3, 3.6, and 3.9 m above ground level) for treatment T1 (1.0 ± 0.5, 0.6 ± 0.3, 0.4 ± 0.2 ng cm^{-2}, respectively) was significantly lower than T2 (3.9 ± 1.0, 2.8 ± 0.9, 2.9 ± 0.9 ng cm^{-2}, respectively) (Figure 9b). However, the sampling height did not significantly affect the aerial deposition for a particular SSCDS treatment. Analysis of coverage samplers indicated that there was negligible mean aerial coverage (<0.1%) for both the treatments.

Table 8. Mean aerial drift losses evaluated for tested treatments.

Off-Target Loss	Treatment	Deposition (ng cm^{-2}) * [Aerial Drift (%)] #	Coverage (%) *
Aerial drift	T1	0.7 ± 0.1 ^{b} [0.02]	0.0 ± 0.0 ^{A}
	T2	3.2 ± 0.4 ^{a} [0.08]	0.0 ± 0.0 ^{A}

* Transformed data was used for statistical analysis; presented data are non-transformed values in (mean ± SEM) format; different lowercase and uppercase letters represent the differences (significant or not significant) in transformed mean at $\alpha = 0.05$; # The values in the square bracket represent the percent of applied active ingredient drifted into the air; The lowercase and uppercase letters in superscript indicates the significant differences in transformed mean at $\alpha = 0.05$.

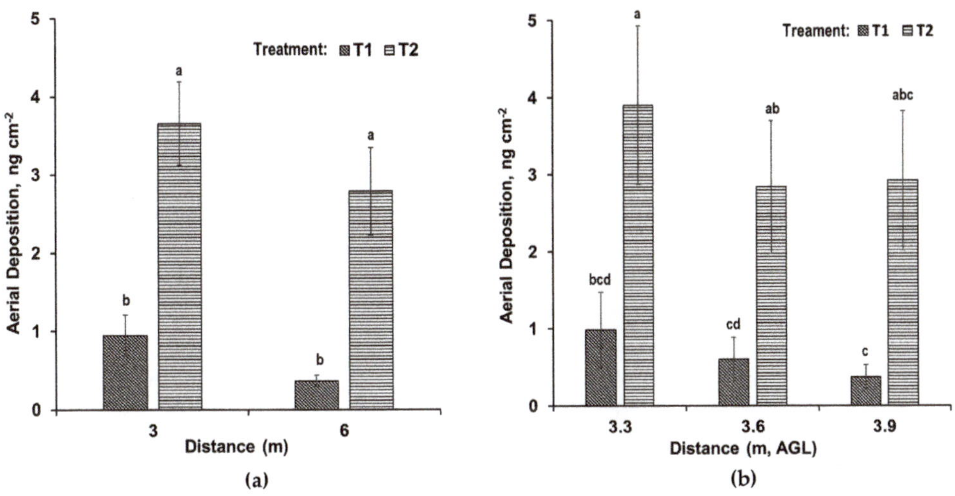

Figure 9. Mean aerial deposition observed at (**a**) 3 m and 6 m downwind and (**b**) 3.3 m, 3.6 m and 3.9 m above ground level for tested SSCDS treatments (i.e., T1 and T2). Different lowercase letters above individual bar plots indicate the differences (significant or not significant) in transformed mean at $\alpha = 0.05$ and associated error bars indicate standard error; presented values are non-transformed mean ground deposition (ng cm^{-2}).

4. Discussion

The study results indicate that the SSCDS configured modified irrigation micro-emitters resulted in comparable canopy deposition and coverage against that of hollow cone nozzles. The modified irrigation micro-emitter configured treatment resulted in 28.0%

and 19.5% higher mean spray deposition and coverage, respectively, compared to hollow cone nozzle configured SSCDS treatment. Despite a considerable numerical difference in deposition and coverage between the modified irrigation micro-emitter and hollow cone nozzle SSCDS treatment, the corresponding difference was statistically non-significant, perhaps due to high variability in deposition and coverage data and a relatively small number of samples. Such variability is very typical in agrochemical application scenarios [20,31,32] and could be reduced by increasing the number of replicates in the experiment. The bottom zone deposition and coverage for both the tested treatments were highest compared to the other sampling zones. The spray droplets that missed the target and spray run-off from the top and mid canopy zone, settled to the bottom, would have resulted in a higher deposition in later zone. The modified irrigation micro-emitter treatment resulted in 59.4% and 33.1% higher bottom zone deposition and coverage, respectively, compared to the hollow cone nozzle treatment. Similarly, mid zone deposition and coverage was higher than the hollow cone nozzle treatment.

Micro-emitter modification also improved the leaf surface deposition and coverage. The abaxial and adaxial deposition of modified irrigation micro-emitter treatment were 38% and 21.6% higher than hollow cone nozzle treatment, respectively. Likewise, the modified treatment had 38.3% and 12.4% higher abaxial and adaxial coverage compared to hollow cone nozzle treatments. Additionally, modified SSCDS treatment resulted in lower CV in spray deposition and coverage between the abaxial and adaxial leaf surfaces. Results indicate that the micro-emitter modification also augmented the leaf level spray uniformity. Previous SSCDS configuration test studies have reported that shower down arrangement with only one nozzle/micro-emitter atop tree canopy was the simplest and the most economical SSCDS configuration [20,22]. However, poor bottom zone and abaxial leaf surface deposition were perceived as a major constraint with such configuration [33]. The presented 3-tier configured SSCDS with modified emitters indicates a substantial increase in the bottom zone and abaxial leaf surface deposition and coverage. Furthermore, the modified SSCDS configuration uses a low-cost micro-emitter (1.5 $/unit) (Table 1) that assisted in reducing the system installation cost by ~12 times, compared to expensive off-the-shelf hollow cone nozzles (20 $/unit) configured SSCDS. Such cost savings is expected to improve its economic viability.

The off-target drift data trends indicate that the modified irrigation micro-emitter treatment (T1) had numerically lower aerial, ground run-off and drift losses compared to treatment configured with hollow cone nozzle (T2). The adjacent mid-row ground deposition and coverage evaluated at 1.5 m downwind were, respectively, 258.5% and 326.5% lower for the modified micro-emitter treatment. Likewise, overall mean mid-row ground deposition and coverage in the modified treatment were 267.3% and 327% lower than hollow cone nozzle treatment. A similar trend was observed for downwind aerial drift. The aerial deposition for T1 recorded at 3 m and 6 m downwind was, respectively, 298% and 373% lower than T2. The overall mean aerial deposition of modified treatment was 395% lower compared to hollow cone nozzle SSCDS treatment. Furthermore, the ground spray coverage at 4.5 m and 7.5 m and aerial coverage at 3 m and 6 m downwind for modified treatment were almost negligible (<0.1%). The hollow cone nozzles produced fine size droplets (Table 3) that are highly susceptible to drift due to longer air suspension time and lighter weight [34–36]. In contrast, the modified micro-emitter produced medium sized droplets (Table 3) that would have reduced the off-target drift potential [37]. Additionally, wind speed during the modified micro-emitter trials was 1.5 times higher than hollow cone nozzle treatment trials (Table 2). With similar wind direction trend during both treatments, higher wind speed would cause higher off-target drift for spray with modified irrigation micro-emitters. Nonetheless, modification in droplet size spectrum would have curtailed pertinent off-target drift losses. This might be one of the reasons why modified micro-emitter configured treatments would have resulted in lesser ground and aerial drift. Therefore, pertinent modification could assist in minimizing the environmental contamination and subsequent hazards [6–8]. Furthermore, the reduced drift for modified

SSCDS treatment would have resulted in improved overall as well as zone and leaf specific canopy deposition and coverage [38].

While a modified irrigation micro-emitter provided a low-cost alternative to the hollow cone nozzles in SSCDS configuration, existing pneumatic spray delivery reservoir drains the residual spray mix onto the soil through auto drain valve for cleaning. During pesticide application, such residues can contaminate the soil [39] and can cause up to 25% pesticide losses to the ground. To overcome this problem, our lab is working on the modification of the existing pneumatic spray delivery reservoirs, which allows partial passage of the air to the nozzle feed line towards the end of the spray cycle so that residues can be sprayed into the canopy instead of draining onto the soil. Pertinent self-cleaning ability would eliminate the need of the auto-drain valve in the reservoir and ground chemical losses.

5. Conclusions

1. The 3-tier SSCDS treatment configured with modified micro-emitters had comparable spray performance with numerically higher spray deposition, coverage and lower off-target drift losses compared to that of a SSCDS configured using off-the-shelf hollow cone nozzles.
2. The modified micro-emitters facilitated the uniform distribution of spray material on upper and lower leaf surfaces. The micro-emitter refinement was thus successful and assisted in improving spray performance.

Our future research will focus on exploring the biological efficacy of the modified SSCDS configuration. These data are needed to support the further development and eventual commercial adaptation of this system.

Author Contributions: L.R.K., R.R. and R.S. provided the idea of this work and contributed to the conceptualization and experiment design. M.L. assisted in emitter design and fabrication. R.R., R.S., L.R.K. and G.-A.H. contributed to the field data collection. R.R. performed the data analysis and prepared the original draft. R.S., L.R.K., G.-A.H., M.G. and M.L. reviewed and revised the manuscript in the current form. Correspondence and request for the paper should be addressed to L.R.K. All authors have read and agreed to the published version of the manuscript.

Funding: This project was funded by the USDA-NIFA Specialty Crop Research Initiative grant program and WNP0745.

Informed Consent Statement: Not applicable.

Acknowledgments: The authors extend their gratitude for Karen Lewis and Bernardita Sallato for allowing us to conduct the trial in their research block. Sincere thanks are also given to Patrick Scharf and Linda Root for their technical and administrative support. We would also like to thank Abhilash Chandel, Jake Schrader, Anura Rathanayake, Haitham Bahlol, Ramesh Sahni, Margaret McCoy, Maia Bloom and Jennifer for their assistance in data collection and analysis.

Conflicts of Interest: The authors declare no conflict of interest.

References

1. USDA; NAAS. Fresh Apples, Grapes, and Pears: World Markets and Trade. 2020. Available online: https://apps.fas.usda.gov/psdonline/circulars/fruit.pdf (accessed on 15 March 2020).
2. USDA; NAAS. Noncitrus Fruits and Nuts. 2019. Available online: https://www.nass.usda.gov/Publications/Todays_Reports/reports/ncit0619.pdf (accessed on 15 March 2020).
3. Fox, R.D.; Derksen, R.C.; Zhu, H.; Brazee, R.D.; Svensson, S.A. A history of air-blast sprayer development and future prospects. *Am. Soc. Agric. Biol. Eng.* **2008**, *51*, 405–410.
4. Sedlar, A.D.; Bugarin, R.M.; Nuyttens, D.; Turan, J.J.; Zoranovic, M.S.; Ponjican, O.O.; Janic, T.V. Quality and efficiency of apple orchard protection affected by sprayer type and application rate. *Span. J. Agric. Res.* **2013**, *11*, 935–944. [CrossRef]
5. EPA. Introduction to Pesticide Drift. 2019. Available online: http://epa.gov/reducing-pesticide-drift/introduction-pesticide-drift (accessed on 10 April 2020).
6. Grella, M.; Gallart, M.; Marucco, P.; Balsari, P.; Gil, E. Ground deposition and airborne spray drift assessment in vineyard and orchard: The influence of environmental variables and sprayer settings. *Sustainability* **2017**, *9*, 728. [CrossRef]

7. Grella, M.; Marucco, P.; Manzone, M.; Gallart, M.; Balsari, P. Effect of sprayer settings on spray drift during pesticide appli-cation in poplar plantations (Populus. spp.). *Sci. Total Environ.* **2017**, *578*, 427–439. [CrossRef] [PubMed]
8. Cunha, J.P.; Chueca, P.; Garcerá, C.; Moltó, E. Risk assessment of pesticide spray drift from citrus applications with air-blast sprayers in Spain. *Crop. Prot.* **2012**, *42*, 116–123. [CrossRef]
9. Whiting, M.D.; Lang, G.; Ophardt, D. Rootstock and training system affect sweet cherry growth, yield, and fruit quality. *HortScience* **2005**, *40*, 582–586. [CrossRef]
10. Ampatzidis, Y.; Whiting, M.D. Training system affects sweet cherry harvest efficiency. *HortScience* **2013**, *48*, 547–555. [CrossRef]
11. Warner, G. Which is Better for Growing Apples, Angled or Upright. Good Fruit Grower. 2014. Available online: https://www.goodfruit.com/which-is-better-for-growing-apples-angled-or-upright/ (accessed on 15 April 2020).
12. Grieshop, M.J.; Emling, J.; Ledebuhr, M. Re-Envisioning Agrichemical Input Delivery: Solid Set Delivery Systems for High Density Fruit Production, Impacts on Off-Target Deposition. In Proceedings of the 2018 ASABE Annual International Meeting, Detroit, MI, USA, 31 July 2018; p. 1.
13. Cross, J.; Walklate, P.; Murray, R.; Richardson, G. Spray deposits and losses in different sized apple trees from an axial fan orchard sprayer: 1. Effects of spray liquid flow rate. *Crop. Prot.* **2001**, *20*, 13–30. [CrossRef]
14. Bode, L.E.; Bretthauer, S.M. Agricultural chemical application technology: A remarkable past and an amazing future. *Am. Soc. Agric. Biol. Eng.* **2008**, *51*, 391–395.
15. Niemann, S.M. Efficacy of a Solid Set Spray System in Modern Apple and Sweet Cherry Orchards. Ph.D. Thesis, Washington State University, Washington, DC, USA, 2014.
16. Klein, R.; Schulze, L.; Ogg, C. Factors Affecting Spray Drift of Pesticides. 2008. Available online: https://www.certifiedcropadviser.org/files/certifications/certified/education/self--study/exam--pdfs/119.pdf (accessed on 30 March 2020).
17. Garcerá, C.; Moltó, E.; Chueca, P. Spray pesticide applications in Mediterranean citrus orchards: Canopy deposition and off-target losses. *Sci. Total Environ.* **2017**, *599*, 1344–1362. [CrossRef] [PubMed]
18. Sinha, R.; Khot, L.R.; Hoheisel, G.-A.; Grieshop, M.J.; Bahlol, H. Feasibility of a Solid set canopy delivery system for efficient agrochemical delivery in vertical shoot position trained vineyards. *Biosyst. Eng.* **2019**, *179*, 59–70. [CrossRef]
19. Niemann, S.M.; Whiting, M.D. Spray coverage in apple and cherry orchards using a solid set canopy delivery system. In *XXIX International Horticultural Congress on Horticulture: Sustaining Lives, Livelihoods and Landscapes (IHC2014)*; International Society for Horticultural Science: Leuven, Belgium, 2014; Volume 1130, pp. 647–654.
20. Sinha, R.; Ranjan, R.; Khot, L.R.; Hoheisel, G.; Grieshop, M.J. Drift potential from a solid set canopy delivery system and an axial–fan air–assisted sprayer during applications in grapevines. *Biosyst. Eng.* **2019**, *188*, 207–216. [CrossRef]
21. Sinha, R.; Ranjan, R.; Khot, L.R.; Hoheisel, G.A.; Grieshop, M.J. Development and performance evaluation of a pneumatic spray delivery based solid set canopy delivery system for spray application in a high–density apple orchard. *Trans. ASABE* **2020**, *63*, 37–48. [CrossRef]
22. Sinha, R.; Khot, L.R.; Hoheisel, G.A.; Grieshop, M.J. Solid set canopy delivery system configured for high-density tall spindle architecture trained apple canopies. *Trans. ASABE* **2020**. revision submitted. [CrossRef]
23. Sinha, R.; Ranjan, R.; Khot, L.R.; Hoheisel, G.; Grieshop, M.J. Comparison of within canopy deposition for a solid set canopy delivery system (SSCDS) and an axial–fan airblast sprayer in a vineyard. *Crop. Prot.* **2020**, *132*, 105124. [CrossRef]
24. Owen-Smith, P.; Perry, R.; Wise, J.; Jamil, R.Z.R.; Gut, L.; Sundin, G.; Grieshop, M.J. Spray coverage and pest management efficacy of a solid set canopy delivery system in high density apples. *Pest Manag. Sci.* **2019**, *75*, 3050–3059. [CrossRef]
25. Owen-Smith, P.; Wise, J.C.; Grieshop, M.J. Season long pest management efficacy and spray characteristics of a solid set canopy delivery system in high density apples. *Insects* **2019**, *10*, 193. [CrossRef]
26. Landers, A.J.; Agnelo, A.; Shayya, W. The Development of a Fixed Spraying System for High-Density Apples. In *Proceedings of the 2006 ASAE Annual Meeting, 9–12 July 2006 (p. 1)*; American Society of Agricultural and Biological Engineers: St. Joseph, MI, USA, 2006.
27. Ranjan, R.; Shi, G.; Sinha, R.; Khot, L.R.; Hoheisel, G.-A.; Grieshop, M.J. Automated solid set canopy delivery system for large-scale spray applications in perennial specialty Crops. *Trans. ASABE* **2019**, *62*, 585–592. [CrossRef]
28. Sharda, A.; Karkee, M.; Zhang, Q.; Ewlanow, I.; Adameit, U.; Brunner, J. Effect of emitter type and mounting configuration on spray coverage for solid set canopy delivery system. *Comput. Electron. Agric.* **2015**, *112*, 184–192. [CrossRef]
29. ASABE S572.3. *Spray Nozzle Classification by Droplet Spectra*; American Society of Agricultural and Biological Engineers: St. Joseph, MI, USA, 2020.
30. R Studio, R Core Team. *R: A language and Environment for Statistical Computing*; R Foundation for Statistical Computing: Vienna, Austria, 2020; Available online: https://www.R-project.org/ (accessed on 15 January 2020).
31. Sinha, R.; Ranjan, R.; Shi, G.; Hoheisel, G.A.; Grieshop, M.; Khot, L.R. Solid set canopy delivery system for efficient agro-chemical delivery in modern architecture apple and grapevine canopies. *Acta Hort.* **2020**, *1269*, 277–286. [CrossRef]
32. Koch, H.; Weisser, P. Sensor equipped orchard spraying–efficacy, savings and drift reduction. *Asp. Appl. Biol.* **2000**, *57*, 357–362.
33. Ranjan, R.; Sinha, R.; Khot, L.R.; Hoheisel, G.A.; Grieshop, M.J.; Ledebuhr, M. Effect of emitter modifications on spray performance of a solid set canopy delivery system in a high-density apple orchard. *Biosyst. Eng.* **2020**. YBENG-D-20-00541, submitted.
34. Potts, S.F. Particle size of insecticides and its relation to application, distribution, and deposit. *J. Econ. Èntomol.* **1946**, *39*, 716–720. [CrossRef] [PubMed]
35. Klingman, G.C.; Noordhoff, L.J. *Weed Control: As a Science*; John Wiley and Sons: Hoboken, NJ, USA, 1961; p. 67.

36. Hofman, V.; Solseng, E. Reducing Spray Drift. 2001. Available online: https://www.ag.ndsu.edu/publications/crops/reducing-spray-drift/ae1210.pdf (accessed on 30 March 2020).
37. Teske, M.E.; Thistle, H.W.; Mickle, R.E. Modeling finer droplet aerial spray drift and deposition. *Appl. Eng. Agric.* **2000**, *16*, 351–357. [CrossRef]
38. Picot, J.J.C.; Kristmanson, D.D.; Basak-Brown, N. Canopy deposit and off-target drift in forestry aerial spraying: The effects of operational parameters. *Trans. ASAE* **1986**, *29*, 90–96. [CrossRef]
39. EPA. Safe Disposal of Pesticides. 2017. Available online: https://www.epa.gov/safepestcontrol/safe-disposal-pesticides (accessed on 30 March 2020).

Article

Assessment of Handler Exposure to Pesticides from Stretcher-Type Power Sprayers in Orchards

Zhinan Wang [1,†], Yuxi Meng [1,2,†], Xiangdong Mei [1], Jun Ning [1], Xiaodong Ma [2,*] and Dongmei She [1,*]

1 State Key Laboratory for Biology of Plant Diseases and Insect Pests, Institute of Plant Protection, Chinese Academy of Agricultural Sciences, No.2 Yuanmingyuan West Road, Beijing 100193, China; 82101182339@caas.cn (Z.W.); myxiixi@cau.edu.cn (Y.M.); xdmei@ippcaas.cn (X.M.); jning@ippcaas.cn (J.N.)

2 College of Science, China Agricultural University, Beijing 100193, China

* Correspondence: dongxm@cau.edu.cn (X.M.); shedongmei@caas.cn (D.S.)

† Zhinan Wang and Yuxi Meng are co-first authors and contributed equally to this work.

Received: 17 November 2020; Accepted: 2 December 2020; Published: 4 December 2020

Abstract: The production and export volume of fruits from China are among the top three in the world. Pesticides are applied to orchards more than 10 times a year to control pests, and stretcher-type power sprayers are widely used to apply chemical pesticides. However, an assessment of pesticide-handler exposure to pesticides in this scenario has not been reported in China. The test pesticide, 30% SYP-9625 concentrate diluted 3000 times, was sprayed on apple orchards in Beijing China. Experiments were conducted to assess dermal and inhalation exposure using standard whole-body dosimetry and air-sampling methodologies. The dermal deposition was the main route of exposure in this study. The dermal unit exposure (UE) of handlers was 350 mg·kg^{-1} a.i. of SYP-9625. The hands accounted for 59% of the total exposure and were the most exposed body part. Inhalation UE was 0.720 mg·kg^{-1} a.i. of SYP-9625 and was negligible compared with dermal exposure. We found that use of protective garments while using stretcher-type powers sprayers reduced dermal pesticide exposure. These results can be used as a reference for the handler's safety in pesticide management and orchard mechanical management.

Keywords: inhalation exposure; dermal exposure; stretcher-type power sprayers; orchards

1. Introduction

While spraying crops, handlers are inevitably exposed to pesticides. Pesticide-handlers are directly exposed to pesticide clouds, allowing pesticides to readily enter the body through the skin and respiratory tract, which may result in acute poisoning and other chronic health issues [1–3]. Therefore, it is necessary to evaluate occupational exposure to pesticides to reduce the degree of exposure and protect handlers [4,5].

Women and the elderly, the main labor force in rural China, are not well educated and have a low level of security awareness [6]. Exposure to pesticides is one of the most serious occupational risks faced by them. Therefore, it is imperative to carry out exposure assessments for pesticides according to the fundamental realities of China.

In recent years, China has gradually paid attention to the occupational exposure assessment of pesticides. Now our exposure assessment is based on the methods used in developed countries in the West. The methods used by different people converge and there are certain differences. Dermal exposure is a key component of pesticide risk assessment and whole-body dosimetry is more accurate in estimating dermal exposure [7,8]. Yang et al. [9] and Gao et al. [10] used the whole-body sampling method to study the exposure of farmers using the 16-type knapsack manual sprayer to spray chlorpyrifos in cornfields. Cao et al. [11] and Chen [12] studied the exposure of farmers to imidacloprid, chlorpyrifos, and lambda-cyhalothrin in wheat fields using whole-body dosimetry.

In 2018, orchards covered 11,875 thousand hectares in China, and fruit output reached 226.88 million tons [13]. The fruits covering the largest area were apples, citrus, and pears and the output of these fruits accounted for 37.6% of the total fruit production in the country [13]. Therefore, it is necessary to assess the exposure risk of operators to pesticides during treatment in apple orchards under defined-use scenarios.

In 1981, Franklin et al. [14] used air monitoring and patch techniques to estimate exposure, proposing an exposure measurement method for orchard application scenarios. Moon and Kim studied the pesticide exposure assessment of the application of fenvalerate and methomyl in apple orchards in Korea [15]. Thouvenin selected insecticide foliar application to a vineyard as the exposure scenario [16]. However, relevant reports on the occupational exposure to pesticides in orchards have not been studied in China.

In developed countries in Europe or North America, the application of pesticide chemicals is very specialized, serialized, and standardized. Depending on the application, each scene has a dedicated application machine [17]. China's pesticide application equipment models are aging, and the market share of manual sprayers reaches 80% [18]. Stretcher-type sprayers are mainly used for fruit-tree applications. This type of sprayer originated in the 1950s and 1960s and was developed to prevent serious injury from rice borers. The structure of the stretcher-type power sprayer consists of five parts—the frame, the power machine, the liquid pump, the water absorption part, and the spray part [19]. The diameter of the droplet is about 400–600 microns [20]. It has a long range and a wide spray coverage and is easy to use, but the pesticide utilization rate is only about 15% [20].

The pesticide SYP-9625 is a broad-spectrum acrylonitrile acaricide that is quick-acting and long-lasting [21]. As a newly-developed acaricide, no studies have been published on exposure risk assessment. The purpose of this work was to study the exposure data of occupational handlers using stretcher-type power sprayers for fruit tree application scenarios.

2. Materials and Methods

2.1. Reagents and Materials

Analytical standard SYP-9625 (CAS No., 1253429-01-4) (98% purity) was purchased from Shenyang Sinochem Agrochemicals R&D Co., Ltd. (Shenyang, China). The Structure of SYP-9625 is in Figure S1. The commercial SYP-9625 formulation used in the field trial was 30% suspension concentrate. Acetone and acetonitrile used for the extraction were of analytical grade; the acetonitrile used for the preparation of standard solutions was of high-performance liquid chromatography grade (Sigma-Aldrich, Steinheim, Germany). Lidded glass jars of 500 mL, 1000 mL, and 2500 mL sizes were used to extract samples. Cotton clothing was obtained from the Evolu Flagship Store (Hangzhou, China).

2.2. Field Trial

Operators were local farmers; they were also volunteers. They were experienced in the application of pesticides and in good health. They had agreed to the experimental procedures and co-operated in this study. Each operator was provided with a full explanation of the study and its requirement, and any potential risks. They could withdraw from the study at any time, and for any reason. A signed, informed consent form was obtained from each operator prior to his/her participation in the study.

The exposure study was conducted in Changping district, Beijing, China. At noon in summer, the temperature is too high to spray pesticide, and it can also cause phytotoxicity to plants. Farmers generally use pesticides in the morning or evening. So our experiments were conducted from 6:30 to 10:00 on 7 June 2017. The temperature was 17.5 to 37 °C, sunny with a relative humidity of 27–87%, and no wind. The experiment included four male handlers (1a, 1b, 2a, 2b, 3a, 3b, 4) with a total of seven exposures. The handlers were approximately 50 years old, with height and weight of

165–175 cm and 60–70 kg, respectively. The apple trees were in the fruiting stage, and their average height was 3 m. The rows were separated by 2.5–3.8 m, with 1.5–2.5 m between trees in the same row.

For pesticide application, the spray solution was prepared by mixing 300 L of water with 100 mL 30% suspension concentrate. Each handler followed their normal pace to spray pesticide across one acre. The application was carried out simultaneously by four stretcher-type electric sprayers. The machinery was provided by the farmers who planted the fruit trees and had been in use for 1 to 4 years. New, old, or different sprayers may have caused different exposures. The sprayer in Figure 1 is a relatively-new stretcher-type electric sprayer. Four farmers had more than ten years of experience in spraying pesticides. They randomly-selected instruments and used them to complete seven application experiments. In the process of application, the handlers sprayed according to their usual spraying habit in order to obtain real exposure. Their spray habits had been formed through long-term work. These farmers' spraying techniques were considered representative of typical spraying behaviors.

Figure 1. Stretcher-type power sprayer.

2.3. *Potential Dermal and Inhalation Exposure Monitoring*

2.3.1. Potential Dermal Exposure Monitoring

Dermal exposure was determined using a whole-body dosimetry method, including the two sets of clothing, inside and outside hat, masks, and inside and outside gloves. The basic requirements were as follows: (1) inner clothing—using a cotton content greater than 70%, white, thin, long-sleeved shirt and trousers, round neck, cuffs, neckline tightened; (2) outer clothing—generally with cotton content greater than 70%, white, thick, long-sleeved shirt and trousers, round neck, cuffs, neckline tightened; (3) inside hat—with eight layers of white gauze (20 × 40 cm); (4) outside hat—single-layer, cotton white hat with brim; (5) masks—medical gauze masks; (6) inner gloves—white thin glove with a cotton content greater than 70%; and (7) outer gloves—white thick-lined gloves with cotton content greater than 70%. For accurate quantitative calculations, the handlers wore two layers of clothing during application, and the inner layer of clothing was used to simulate the skin.

2.3.2. Potential Inhalation Exposure Monitoring

Inhalation exposure was measured using a personal air monitor equipped with a portable battery-operated sampling pump and a solid sorbent tube (ORBO 609 Amberlite® XAD-2 400/200 mg). XAD-2 resin was used for capturing the pesticides in the air (each handler's breathing zone). Personal air sampling pumps with XAD-2 filter tubes were used to monitor inhalation exposure. Inhalation UE was calculated with the formula:

$$\mathrm{UE}(\mathrm{mg{\cdot}kg^{-1} of\ a.i.handled}) = \frac{\mathrm{exposure\ dose(mg)} \times \mathrm{application\ time(min)} \times \mathrm{air\ change\ rate(L\ min^{-1})}}{\mathrm{flow\ rate}(\mathrm{L\ min^{-1}}) \times \mathrm{sampling\ time(min)} \times \mathrm{kg\ of\ a.i.\ handled}}$$

An air change rate of 29 L·min^{-1} and an air sampler's pumping rate of 2 L·min^{-1} was assumed [22]. The application and sampling time was assumed to be equal because the orchard size was small and there was no need for rest periods during applications.

2.4. Handler Sampling

After the end of the application, the air sampling pump was turned off and the XAD-2 tubes were collected for cryopreservation and transportation. Sampling time was defined as the time from the start to the end of spraying. The protective clothing was divided into six parts as depicted in Figure 2, labeled, and transported in the dark to be stored under freezing conditions.

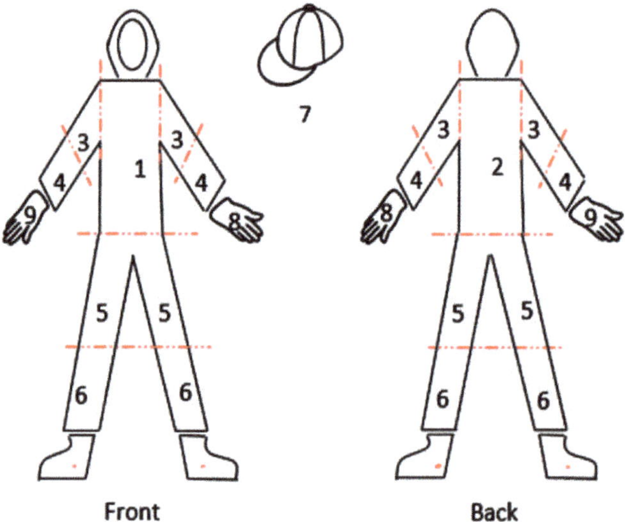

Figure 2. Garment sectioning for whole-body analysis.

Some body parts were cleaned and the resultant wash was collected, as follows: the assistant wore clean disposable gloves and thoroughly wiped the worker's face/neck with moist medical gauze. Approximately 4 mL of 0.01% Aerosol OT(Sodium dioctyl sulfosuccinate) was evenly distributed over the gauze. The steps above were repeated. Each gauze was then collected and labeled. The handlers immersed both hands in 400 mL of 0.01% Aerosol OT solution and scrubbed carefully for at least 30 s. The hands were then rinsed with about 100 mL of solution, which was consistent with the above. This 500 mL sample was collected and labeled [23].

To evaluate the stability of SYP-9625 during storage and transportation, an on-site addition recovery test was carried out in the area near the application site. A certain concentration of pesticide standard solution was prepared, the pesticide solution was spread evenly on a complete part of the clothes (such as thighs), and the volume recorded. Then the clothes were put in a sealed bag and transported under the same conditions as the experimental samples. Each material had a different amount of additive, and two repetitions were taken: (1) underwear (size: 30 cm × 30 cm) 10× and 100× LOQ (limit of quantitation); (2) outer coat (size: 30 cm × 30 cm) 100× and 1000× LOQ; (3) handwashing solution (solution volume: 50 mL) 10× and 100× LOQ; (4) facial and neck wipes 20× and 200× LOQ;

(5) inner gloves 10× and 100× LOQ; (6) outer gloves 100× and 1000× LOQ; and (7) air filter samples 10× and 100× LOQ.

The LOQ of each component and matrix reflected the minimum amount of data that could be quantified.

The samples from the on-site additional recovery test had the same environmental conditions as the samples from the application and used the same transport and storage conditions.

2.5. Chromatographic Conditions

SYP-9625 was analyzed using Ultra-performance liquid chromatography tandem mass spectrometry (UPLC-MS) (AQUITY TQD, Waters, Milford, MA, USA) with a C18 column (2.1 mm × 100 mm, 1.7 μm, Waters, UPLC® BEH). The mobile phase was acetonitrile (A) and 0.1% formic acid-water (B). The linear mobile phase gradient started at 40% A (0 to 0.5 min), increasing to 80% A (0.5 to 3.5 min), after which the column was equilibrated at 40% A (3.5 to 5.0 min). Tandem mass spectrometry was operated in positive electrospray ionization (ESI) and multiple reaction monitoring (MRM) mode. The optimized values of the MS parameters were as follows: source temperature 150 °C, desolvation temperature 350 °C, cone gas (N^2) flow of 50 $L{\cdot}h^{-1}$, desolvation gas (N^2) flow of 650 $L{\cdot}h^{-1}$. The analytical instrument control, data acquisition, and processing were performed by MassLynx V4.1 software (Waters). The mass spectrometric parameters for UPLC-MS determination of SYP-9625 concentrations were listed in Table S1. The UPLC-MS chromatogram for SYP-9625 were in Figure S2.

2.6. Linear Range

Linear regression was performed between 0.0014 to 1.4 $mg{\cdot}L^{-1}$ for SYP-9625. The derived calibration curve formula was $y = 241643x + 4475.8$ ($R^2 = 0.9987$). The limit of detection (LOD) of the compounds was calculated on a signal-to-noise (S/N) ratio of 3 concerning the background noise obtained from the blank sample, whereas the limit of quantitation (LOQ) was via an S/N ratio of 10. The LOD of SYP-9625 was 0.010 $ug{\cdot}L^{-1}$ and LOQ was 0.030 $ug{\cdot}L^{-1}$.

2.7. Extraction and Recovery of SYP-9625

Pesticide extraction methods for similar matrices have been reported in the literature, and the methods are briefly described as follows [23]: According to the size of different parts of the clothes, the protective suits were ultrasonically extracted with different volumes of acetone for 20 min at 25 °C. The operation was repeated once for better recovery. The volume of the extraction solvent was sufficient to immerse the clothing and was recorded in detail. The extract was then concentrated on a rotary evaporator and dissolved in 2 mL EtOAc(Ethyl acetate). The XAD-2 sample was extracted twice with 5 mL of acetone, then concentrated with a nitrogen evaporator and dissolved in 1 mL of acetonitrile. For 0.01% Aerosol OT, 10 mL of acetonitrile was added to a 10 mL sample and shaken vigorously, then NaCl was added to the mixture, shaken, and centrifuged at 3600 $r{\cdot}min^{-1}$ for 5 min to separate the organic solvent from the water. Finally, 1 mL of the resulting supernatant was directly injected into UPLC-MS with an injection volume of 10 uL.

2.8. Statistical Analysis

Results were expressed as mean ± SD and statistical significance was determined by one-way analysis of variance (ANOVA) followed by the least significant difference (LSD) using the statistical software SPSS (SPSS 19.0; SPSS Inc., Chicago, IL, USA). The level of statistical significance was established at $p < 0.05$.

3. Results and Discussion

3.1. Validation of the Analytical Method

Recovery experiments were carried out at three different fortification levels in nine replicates and the relative recovery rates were calculated via the matrix-matched calibration curves. Acetone was chosen as the extraction solvent for cotton wool, gauze, and XAD-2. In this study, SYP-9625 detected in different materials had a high recovery rate, and a method of detection and analysis of SYP-9625 concentrations in different matrices was established. Acceptable recoveries were obtained in the ranges of 91.02–102.35% from cotton wool, 97.34–103.45% from gauze, 95.33–107.26% from outer gloves, 93.28.9–97.65% from XAD-2, and 103.25–110.37% from 0.01% Aerosol OT. The recovery and precision of SYP-9625 (expressed as relative standard deviation) are shown in Table 1, RSDs were less than 10% in all samples, demonstrating the excellent repeatability of the process. Therefore, a combination of UPLC-MS detection, and the utilized extraction method could serve as a conventional detection method for SYP-9625 in all of the above matrices.

Table 1. SYP-9625 recovery from field-fortified samples.

Material	Fortification Concentration ($mg{\cdot}kg^{-1}$)	Recovery (%)	RSD (%)
	0.1	97.28	2.8
Cotton clothing	0.2	91.02	3.1
	1	102.35	3.4
	0.01	101.00	2.0
Gauze	0.02	103.45	2.6
	0.1	97.34	1.8
	0.05	95.33	4.7
Outside gloves	0.1	107.26	3.6
	0.2	105.15	5.3
	0.01	94.84	3.2
XAD-2	0.02	97.65	1.9
	0.04	93.28	2.8
	0.025	110.37	2.2
0.01% Aerosol OT	0.05	106.00	1.6
	0.1	103.25	3.8

Since this orchard is far away from our laboratory, we needed to clarify whether the pesticides on the various exposed substrates would be lost during transportation The recovery rate of the sample added to the on-site recovery test was approximately 91.5%. This proves that the loss of pesticides during storage and transportation is below 10%. The OECD(Organization for Economic Co-operation and Development) pointed out that the samples were sufficiently stable within 30% decline of the recovery rate [24].

3.2. Dermal and Inhalation Exposure during Application

In this study, we measured pesticide exposure by dermal and inhalation routes using standard systemic dosimetry and air sampling methods. In the practical application of pesticides, farmers wear clothes with at least one layer of clothing. Actual dermal exposure (ADE) is the amount of pesticide that passes through the clothes and is exposed to the skin. The handlers wear two layers of clothing during application and the inner layer of clothing is used to simulate the skin. Potential dermal exposure (PDE) during application is the sum of dosimeter readings for the inner and outer layers. The pesticide test results for the inner clothing, hand wash, and face/neck wipes were calculated as actual dermal exposure (ADE) [23]. The inhalation exposure was calculated as the amount of active compound adsorbed by the XAD-2 sorbent tube on the air sampler throughout the application period. The pesticide exposure in this study was measured by PDE. Table 2 lists the pesticide unit exposures of the two layers of clothing inside and outside the applicator. According to the literature, the occupational exposure of pesticides in the field is quite different for different pesticide application personnel, and the difference is related to the operating habits and proficiency [10]. Therefore, the final experimental results are based on average.

Table 2. Unit exposures at various parts of the body (mg/kg a.i. handled).

Part of Garment	1	2	3	4	5	6	7	Mean	SD
a1	0.10	0.18	0.32	0.07	0.09	0.16	0.12	0.15	0.08
a2	0.23	0.18	0.08	0.08	0.06	0.08	0.06	0.11	0.07
a3	0.07	0.07	0.02	0.11	0.66	0.02	0.01	0.14	0.23
a4	0.02	0.57	0.05	0.05	0.03	0.05	0.09	0.12	0.20
a5	0.18	0.23	0.14	0.05	0.25	0.08	0.13	0.15	0.07
a6	0.15	0.92	0.12	0.30	0.10	0.02	0.16	0.25	0.31
a7	0.78	0.26	0.13	0.14	1.23	2.58	4.03	1.31	1.48
a8	0.98	99.48	7.43	0.22	0.17	2.35	53.78	23.49	38.71
a9	0.31	124.26	9.27	0.25	0.02	2.63	40.19	25.28	45.97
b1	44.18	31.01	4.92	6.76	3.75	17.83	18.46	18.13	15.03
b2	7.07	5.54	4.56	4.64	4.28	42.70	16.04	12.12	14.11
b3	4.87	14.36	2.50	3.71	4.85	21.38	15.53	9.60	7.38
b4	4.12	47.16	5.02	6.59	4.57	14.92	33.84	16.60	17.17
b5	11.41	20.11	11.32	0.04	3.95	22.07	28.13	13.86	10.10
b6	66.30	98.75	135.68	0.08	7.81	42.49	106.29	65.34	51.37
b7	2.52	9.39	0.58	0.26	4.96	10.91	16.13	6.39	5.95
b8	35.95	253.94	59.09	17.15	32.62	86.76	128.87	87.77	82.50
b9	16.12	234.01	23.58	31.48	9.82	91.44	72.15	68.37	79.07
10	1.33	0.26	0.21	0.12	0.29	0.38	2.33	0.70	0.83
11	0.16	0.05	0.01	0.02	0.02	0.19	0.64	0.16	0.23
12	0.31	0.06	0.15	0.02	0.02	0.20	0.59	0.19	0.20
13	0.03	0.14	0.02	0.01	0.02	0.02	0.11	0.05	0.05
14	0.009	0.005	0.01	0.004	0.048	0.02	0.04	0.02	0.02

A, inner-layer garment; b, outer-layer garment; 1, front torso (above the waist); 2, rear torso (above the waist); 3, right and left upper arms (shoulder to elbow); 4, right and left forearms (elbow to cuff); 5, right and left thighs (waist to knee); 6, right and left lower shins (knee to cuff); 7, cap; 8, left glove; 9, right glove; 10, mask; 11, face wipe; 12, neck wipe; 13, hand washes; 14, XAD-2.

3.2.1. Dermal Unit Exposure during Application

The total PDE of handlers was 350 mg·kg^{-1}. These data is close to the exposure assessment data for the fruit tree application scenario reported by Zhao et al. [25] and lower than the exposure data for the knapsack sprayer in cornfields and peanut fields reported by Gao et al. [10] and Chen et al. [12], but higher than the exposure data using pesticide speed sprayer application in an apple orchard [26]. As shown in Table 3, according to the UE of each garment, the highest contaminated sections were hands and shins, accounting for an average of 59% and 19% of total dermal exposure, respectively.

Table 3. Distribution of dermal exposure during application (mg/kg a.i. handled).

	Exposure Levels of ADE	Total Exposure Levels of PDE
Front torso	0.15 ± 0.08b	18.13 ± 15.03a
Rear torso	0.11 ± 0.07b	12.12 ± 14.11a
Upper arms	0.14 ± 0.23b	9.60 ± 7.38a
Forearms	0.12 ± 0.20b	16.60 ± 17.17a
Thighs	0.15 ± 0.07b	13.86 ± 10.10a
Shins	0.25 ± 0.31b	65.34 ± 51.37a
Hands	48.82 ± 84.21a	204.96 ± 241.57a
Head	1.31 ± 1.48b	8.75 ± 8.17a
Total	51.05 ± 84.70b	350.28 ± 306.97a

All data are presented as mean ± standard error (n = 7). Different lessters(a, b) in each row indicate significant difference at $p \leq 0.05$ (least significant difference).

We found that the exposure of hands, which refers to the sum of the exposure of the left and right gloves plus that of the handwashing solution, was considerably greater than other body parts. This is different from the results of Zhao et al. [15,24]. While Li et al. [27] showed that when spraying high places with a spray gun, the most contaminated sections were hands. This could be attributed to the different types of application equipment used. In our experiments, the applicator needed two hands to hold the sprayer in order to work properly, and the hands were the only body part in direct contact with the applicator. High pressure can cause pesticide liquid overflow in the connection between the hose and spray lance onto the hand. Further, most of the pesticide liquid sprayed from the nozzle, but liquid flow along the spray lance to the hand was possible. In addition, any problems that occurred during the application process needed to be addressed using both hands, such as pulling a hose for pesticide delivery, and handling the leakage of pesticide [28,29]. The high exposure level of shins may be due to a large number of pesticides deposited on the ground weeds during the application of pesticides into the base of apple trees [30,31]. In addition, handlers needed to walk through the weeds in order to facilitate the application of pesticides [25]. These results are similar to those obtained by Noh et al. [32], who reported when spraying onto lower crops (about 80–100 cm high), workers are exposed while moving.

However, when handlers are wearing long pants and pure cotton gloves, hand exposure decreased by 76%, shin exposure decreased more than 99%, and the ADE was reduced by 85.43% to 51.1 mg·kg^{-1}. This shows that clothing plays a very good role in protecting against pesticides, especially for those who regularly apply pesticides. This is consistent with the conclusions of Ren et al. [33] and An et al. [34]. Hands were still exposed significantly, so the protection of the hands during the application of pesticides should be given extra attention. In order to further reduce the contamination of the hands, we recommend that handlers wear impermeable gloves such as chemical-resistant gloves. The exposure of the left and right hand is approximately the same (Table 4), because during the application process, the handlers need to change the way in which the spray gun is held based on the direction in which they are walking, and the height of the fruit tree, in order to facilitate spraying.

Table 4. Exposure levels of hands (mg/kg a.i. handled).

	ADE of Hands	PDE of Hands
Left hands	23.49 ± 38.71a	111.26 ± 120.23a
Right hands	25.28 ± 45.97a	93.61 ± 123.10a

All data are presented as mean ± standard error (n = 7). Letter (a) in each column indicate significant a difference at $p \leq 0.05$ (least significant difference).

In the orchard application scenario, a handler stands under the tree to apply the pesticide to the oblique crown above the tree. The liquid droplets collide with the leaves and bounce to fall under the force of gravity [35]. The pesticide drifts to the head and body of the handler. During our experiment, the handler's head was protected by a cotton baseball cap. The exposure to the head was not significant, at only 8.75 mg·kg^{-1}, which may be related to the small overall head area. As can be seen from Table 3, the wearing of a cap reduced the skin exposure of the head to 1.31 mg·kg^{-1}, and the protection rate was about 85%. However this protection rate was relatively low compared with other body parts. Since the brain is an essential organ that controls many human organs, the head should be better protected. Therefore, we recommend wearing a wide-brimmed sun visor with cotton or waterproof material during the actual application of pesticides. This can increase the coverage of the head and neck to prevent the pesticide liquid from scattering on the head. At the same time, it can also protect the skin from sunlight and reduce the damage of ultraviolet rays to the skin. In fact, farmers usually wear straw hats when they work. This kind of hat can prevent pesticides from falling on the face and neck. This protective measure can be considered in future experiments.

3.2.2. Inhalation Exposure during Application

Inhalation exposure occurred when airborne pesticide vapors or droplets appeared in working areas owing to the application of pesticides.

In this application scenario, the inhalation unit exposure was 0.72 mg·kg^{-1}, which was only 0.13% of the total skin exposure. The data were slightly higher than the orchard pesticide application scenario reported in Korea (0.7×10^{-3} mg) [25]. This may be because our fruit trees are denser, the workers are close to the fruit trees during the application process, and the whole process of application is under the cloud of pesticides, therefore more pesticides are inhaled.

Stretcher-type electric sprayers are widely used in orchards. Therefore, more experimental data on citrus, grape, pear, and other fruits are needed to get diverse results, also a larger sample size would be good for representative results.

4. Conclusions

This occupational exposure assessment studied the application scenario of a stretcher-type electric sprayer, using the apple tree as an example. This is a typical Chinese orchard application scenario. This is the first time a study has focused on occupational pesticide exposure in orchards in China. The data from this study are a valuable guide for the application of pesticides on fruit trees such as citrus, pear, peach, and other small shrub type fruit trees.

In this study, we performed a systematic occupational exposure assessment on stretcher-type electric sprayer application of SYP-9625 in Chinese orchard fields using standard whole-body dosimetry and air sampling methodologies. We also verified the method of extracting and detecting SYP-9625 from various matrix materials. The results indicate that the hand is the most exposed body part and needs special protection. However, the total exposure is at the same level as that shown in other countries' research. One layer of clothing could protect from dermal exposure, causing a reduction of approximately 86%. Therefore, long trousers and waterproof gloves are necessary to protect the health and safety of the handler. The stretcher-type electric sprayer is China's most important orchard application equipment, but its related occupational exposure data are scarce. Therefore, it is necessary to strengthen the data in this research area.

Our experiment was only carried out in Beijing, and more studies should be performed to obtain extensive experimental results in the future. Our data might help establish accurate and strategic predictive exposure models and databases for risk assessment and pesticide registration.

Supplementary Materials: The following are available online at http://www.mdpi.com/2076-3417/10/23/8684/s1, Table S1: Mass spectrometric parameters for UPLC-MS determination of SYP-9625 concentrations, Figure S1: the structure of SYP-9625, Figure S2: the UPLC-MS chromatogram for SYP-9625.

Author Contributions: Experimental design, Z.W. and D.S.; investigation, Z.W. and Y.M.; data analysis, Z.W. and Y.M.; writing—original draft, Z.W. and Y.M.; writing—review and editing, X.M. (Xiangdong Mei) and D.S.; funding acquisition, D.S. and X.M. (Xiaodong Ma); supervision, J.N. All authors have read and agreed to the published version of the manuscript.

Funding: This research was funded by The National Key Research and Development Program of China (grant number 2016YFD0200201), the National Natural Science Foundation of China (grant number 31772175), and the National Natural Science Foundation of China (grant number 31621064).

Conflicts of Interest: The authors declare no conflict of interest.

References

1. Damalas, C.; Koutroubas, S. Farmers' exposure to pesticides: Toxicity types and ways of prevention. *Toxics* **2016**, *4*, 1. [CrossRef] [PubMed]
2. Baharuddin, M.R.B.; Sahid, I.B.; Noor, M.A.B.M.; Sulaiman, N.; Othman, F. Pesticide risk assessment: A study on inhalation and dermal exposure to 2,4-D and paraquat among Malaysian paddy farmers. *J. Environ. Sci. Health Part B* **2011**, *46*, 600–607. [CrossRef] [PubMed]

3. Verger, P.J.P.; Boobis, A.R. Reevaluate pesticides for food security and safety. *Science* **2013**, *341*, 717–718. [CrossRef] [PubMed]
4. Blanco, L.E.; Aragn, A.; Lundberg, I.; Lidn, C.; Wesseling, C.; Nise, G. Determinants of dermal exposure among Nicaraguan subsistence farmers during pesticide applications with backpack sprayers. *Ann. Occup. Hyg.* **2005**, *49*, 17–24. [CrossRef] [PubMed]
5. Wei, Q.; Tao, C.; Song, W.; Qu, M. Pesticide risk assessment and its status quo and countermeasures. *Qual. Saf. Agro-Prod.* **2010**, 41–45. [CrossRef]
6. National Bureau of Statistics. Communiqué on Major Data of the Third National Agricultural Census (No. 5). 2017. Available online: http://www.stats.gov.cn/tjsj/tjgb/nypcgb/qgnypcgb/201712/t20171215_1563599.html (accessed on 17 November 2020).
7. Großkopf, C.; Mielke, H.; Westphal, D.; Erdtmann-Vourliotis, M.; Hamey, P.; Bouneb, F.; Rautmann, D.; Stauber, F.; Wicke, H.; Maasfeld, W.; et al. A new model for the prediction of agricultural operator exposure during professional application of plant protection products in outdoor crops. *J. Verbrauch. Lebensm.* **2013**, *8*, 143–153. [CrossRef]
8. Abukari, W. Pesticides applicator exposure assessment: A comparison between modeling and actual measurement. *J. Environ. Earth Sci.* **2015**, *5*, 101–115.
9. Yang, F.; Kai, W.; Zhang, W.; Liu, F. Using the protective clothing to simulate the body dermal to measure the exposure of the operators in corn field. *J. Agro-Environ. Sci.* **2013**, *32*, 1979–1983. [CrossRef]
10. Gao, B.; Tao, C.; Ye, J.; Ning, J.; Mei, X.; Jiang, Z.; Chen, S.; She, D. Measurement of operator exposure to chlorpyrifos. *Pest Manag. Sci.* **2013**, *70*, 636–641. [CrossRef]
11. Cao, L.; Chen, B.; Zheng, L.; Wang, D.; Liu, F.; Huang, Q. Assessment of potential dermal and inhalation exposure of workers to the insecticide imidacloprid using whole-body dosimetry in China. *J. Environ. Sci.* **2015**, *27*, 139–146. [CrossRef]
12. Chen, B. *Preliminary Evaluation of Pesticide Exposure of Pesticides in Wheat Field and Peanut Field Spray Operations*; Shandong Agricultural University: Taian, China, 2012. [CrossRef]
13. National Bureau of Statistics. Available online: https://data.stats.gov.cn/easyquery.htm?cn=C01 (accessed on 17 November 2020).
14. Franklin, C.A.; Fenske, R.A.; Greenhalgh, R.; Mathieu, L.; Denley, H.V.; Leffingwell, J.T.; Spear, R.C. Correlation of urinary pesticide metabolite excretion with estimated dermal contact in the course of occupational exposure to Guthion. *J. Toxicol. Environ. Health* **1981**, *7*, 715–731. [CrossRef]
15. Kim, E.; Moon, J.K.; Choi, H.; Hong, S.M.; Lee, D.H.; Lee, H.; Kim, J.H. Exposure and risk assessment of insecticide methomyl for applicator during treatment on apple orchard. *J. Korean Soc. Appl. Biol. Chem.* **2013**, *56*, 123. [CrossRef]
16. Thouvenin, I.; Bouneb, F.; Mercier, T. Operator dermal exposure and protection provided by personal protective equipment and working coveralls during mixing/loading, application and sprayer cleaning in vineyards. *Int. J. Occup. Saf. Ergon.* **2017**, *23*, 229–239. [CrossRef] [PubMed]
17. Liu, F.; Zhang, X.; Ma, W.; Liu, X. Transactions of the Chinese society of agricultural machinery. *J. Agric. Mech. Res.* **2010**, *32*, 246–248. [CrossRef]
18. National Agro-Tech Center Pesticide and Medicine Division. Overview of Pest Control in 2017 and Analysis of Pesticide Equipment Market in 2018. 2017, Volume 30, pp. 30–31. Available online: https://kns.cnki.net/KCMS/detail/detail.aspx?dbcode=CJFDdbname=CJFDLAST2018filename=NYSC201730023uid=WEEvREcwSlJHSldSdmVqM1BLUWh5QjR4OTZGL0xpdFM5dHNLRHh5SmJNWT0=$9A4hF_YAuvQ5obgVAqNKPCYcEjKensW4IQMovwHtwkF4VYPoHbKxJw!!v=MDM0MzFQcjQ5SFo0UjhlWDFMdXhZUzdEaDFUM3FUcldNMUZyQ1VSN3FlWmVac0ZDbm5WTHJMS3pUWWJiRzRIOWI= (accessed on 17 November 2020).
19. Zhou, Q. The use and maintenance of stretcher-type power sprayer and duster. *J. Rural Best Knowl.* **2016**, *601*, 49–50.
20. Research and Application of the Implementation of Zero Growth Strategy of Pesticide Use-Six Projects. Available online: http://www.chinapesticide.org.cn/ddswny/1078.jhtml (accessed on 17 November 2020).
21. Ouyang, J.; Tian, Y.; Jiang, C.; Yang, Q.; Wang, H.; Li, Q.; Gao, Y. Laboratory assays on the effects of a novel acaricide, SYP-9625 on *Tetranychus cinnabarinus*(Boisduval) and its natural enemy, *Neoseiulus californicus*(McGregor). *PLoS ONE* **2018**, *13*, e0199269. [CrossRef] [PubMed]
22. USEPA. Pesticide Assessment Guidelines, Subdivision U, Applicator Exposure. 1987. Available online: http://www.cdpr.ca.gov/docs/whs/memo/hsm98014 (accessed on 17 November 2020).

23. Ren, J.; Tao, C.; Zhang, L.; Ning, J.; Mei, X.; She, D. Potential exposure to clothianidin and risk assessment of manual users of treated soil. *Pest. Manag. Sci.* **2017**, *73*, 1798–1803. [CrossRef]
24. OECD. *Test No 506: Stability of Pesticide Residues in Stored Commodities*; OECD Publishing: Paris, France, 2007. [CrossRef]
25. Zhao, M.A.; Yu, A.; Zhu, Y.Z.; Kim, J.H. Potential dermal exposure to flonicamid and risk assessment of applicators during treatment in apple orchards. *J. Occup. Environ. Hyg.* **2015**, *12*, D147–D152. [CrossRef]
26. Moon, J.; Lee, J.; Maasfeld, W.; Kim, J.; Kim, E.; Lee, J.; Shin, Y.; Lee, J.; Choi, H. Whole body dosimetry and risk assessment of agricultural operator exposure to the fungicide kresoxim-methyl in apple orchards. *Ecotoxicol. Environ. Saf.* **2018**, *155*, 94–100. [CrossRef]
27. Li, Z.; Liu, W.; Wu, C.; She, D. Effect of spraying direction on the exposure to handlers with hand-pumped knapsack sprayer in maize field. *Ecotoxicol. Environ. Saf.* **2019**, *170*, 107–111. [CrossRef]
28. Cao, L.; Zhang, H.; Li, F.; Zhou, Z.; Wang, W.; Ma, D.; Yang, L.; Zhou, P.; Huang, Q. Potential dermal and inhalation exposure to imidacloprid and risk assessment among applicators during treatment in cotton field in China. *Sci. Total Environ.* **2018**, *624*, 1195–1201. [CrossRef] [PubMed]
29. Landers, A. Protecting the operator-are we making an impact? *Asp. Appl. Biol.* **2004**, *71*, 357–364.
30. Hughes, E.A.; Zalts, A.; Ojeda, J.J.; Flores, A.P.; Glass, R.C.; Montserrat, J.M. Analytical method for assessing potential dermal exposure to captan, using whole body dosimetry, in small vegetable production units in Argentina. *Pest. Manag. Sci.* **2006**, *62*, 811–818. [CrossRef] [PubMed]
31. An, X.; Ji, X.; Jiang, J.; Wang, Y.; Wu, C.; Zhao, X. Potential dermal exposure and risk assessment for applicators of chlorothalonil and chlorpyrifos in cucumber greenhouses in China. *Hum. Ecol. Risk Assess.* **2015**, *21*, 972–985. [CrossRef]
32. Noh, H.H.; Lee, J.Y.; Park, H.K.; Lee, J.W.; Jo, S.H.; Kim, J.H.; Kwon, H.; Kyung, K.S. Risk of dermal and inhalation exposure to chlorantraniliprole assessed by using whole-body dosimetry in Korea. *Pest. Manag. Sci.* **2019**, *75*, 1159–1165. [CrossRef]
33. Ren, J.; Li, Z.; Tao, C.; Zhang, L.; Zhao, H.; Wu, C.; She, D. Exposure assessment of operators to clothianidin when using knapsack electric sprayers in greenhouses. *Int. J. Environ. Sci. Technol.* **2019**, *16*, 1471–1478. [CrossRef]
34. An, X.; Wu, S.; Guan, W.; Lv, L.; Liu, X.; Zhang, W.; Zhao, X.; Cai, L. Effects of different protective clothing for reducing body exposure to chlorothalonil during application in cucumber greenhouses. *Hum. Ecol. Risk Assess.* **2018**, *24*, 14–25. [CrossRef]
35. Lee, J.Y.; Noh, H.H.; Park, H.K.; Jeong, H.R.; Jin, M.J.; Park, K.H.; Kim, J.H.; Kyung, K.S. Exposure Assessment of Apple Orchard Workers to the Insecticide Imidacloprid Using Whole Body Dosimetry During Mixing/Loading and Application. *Korean J. Pestic. Sci.* **2016**, *20*, 271–279. [CrossRef]

Article

Nonlethal Effects of Pesticides on Web-Building Spiders Might Account for Rapid Mosquito Population Rebound after Spray Application

Stefan N. Rhoades and Philip K. Stoddard *

Institute of Environment, Department of Biological Sciences, Florida International University, 11200 SW 8th St., Miami, FL 33199, USA; stefannrhoades@gmail.com
* Correspondence: stoddard@fiu.edu; Tel.: +001-305-342-0161

Featured Application: In this study, a broad-spectrum insecticide is shown to halt mosquito capture by orb-weaving spiders, even when the application does not kill the spiders. Reduced prey-capture, even temporary, can allow mosquito populations to rebound quickly. Adoption of other mosquito control methods, such as bacterial larvicides, avoids these potential problems.

Abstract: Spiders are important population regulators of insect pests that spread human disease and damage crops. Nonlethal pesticide exposure is known to affect behavior of arthropods. For spiders such effects include the inability to repair their webs or capture prey. In this study, nonlethal exposure of Mabel's orchard spider (*Leucauge argyrobapta*) to the synthetic pyrethroid permethrin, via web application, interfered with web reconstruction and mosquito capture ability for 1–3 days. The timing of this loss-of-predator ecosystem function corresponds to the rapid population rebound of the yellow fever mosquito (*Aedes aegypti*) following insecticide application to control arbovirus epidemics. We suggest this temporal association is functional and propose that follow-up study be conducted to evaluate its significance.

Keywords: *Aedes aegypti*; *Leucauge argyrobapta*; nontarget effects; permethrin; pyrethroid; Tetragnathidae

Citation: Rhoades, S.N.; Stoddard, P.K. Nonlethal Effects of Pesticides on Web-Building Spiders Might Account for Rapid Mosquito Population Rebound after Spray Application. *Appl. Sci.* **2021**, *11*, 1360. https://doi.org/10.3390/app11041360

Academic Editor: Giuseppe Manetto
Received: 17 December 2020
Accepted: 31 January 2021
Published: 3 February 2021

Publisher's Note: MDPI stays neutral with regard to jurisdictional claims in published maps and institutional affiliations.

1. Introduction

Non-target mosquito predators, including spiders and wasps, are often harmed by neurotoxic pesticides [1]. Orchards where conventional insecticides are used have far lower spider populations and species counts than unsprayed orchards [2]. Spiders with bodies larger than mosquitoes are not killed by ultra-low volume (ULV) spraying of insecticides such as permethrin [3]. However, studies such as these sometimes conflate survival with lack of harm, and they have not examined effects on small-bodied spiders, which do capture mosquitoes. Spiders exposed to sublethal pesticides change their behavior in ways that reduce prey capture. For instance, sublethal pyrethroid spray residues lower activity rates of spiders, even when the spiders contacted the pyrethroids on foliage 20 days after application [4]. Following nonlethal exposure to Spinosad, an acetylcholine disrupter, the orb-weaving spider *Agalenatea redii* showed irregularities in web design and lower prey capture activity [5]. An increase in spider migration was found within a large plantation where sublethal amounts of organophosphate pesticides were sprayed around its borders, suggesting spiders sense and actively avoid areas with some insecticide treatments [6].

Spider webs are particularly effective at capturing insecticide sprays, retaining an order of magnitude higher concentration than an equivalent area of paper [7]. Orb-weaving spiders recycle their webs daily, ingesting the spiral silk strands while leaving some of the radial structural strands intact as a scaffold for rebuilding [8–11]. By consuming the web, orb-weaving spiders ingest higher amounts of sprayed pesticides than would contact their bodies and legs. Araneidae typically construct orb webs in a vertical plane, whereas Tetragnathidae typically construct webs in the horizontal plane [12]. Horizontal webs should intercept more insecticide droplets following a spray application, making tetragnathids particularly vulnerable.

Following aerial insecticide application in New Orleans, yellow fever mosquitoes (*Aedes aegypti*) rebounded faster than the population model predicted [13]. The authors attributed this rapid increase to release from larval competition, a phenomenon well-documented in the literature [14–16]. However, the data presented in that study show that the rebound occurred within the 3-day non-feeding window prior to eclosure, a period when there can be no competition for food. A similar effect could be seen in Miami's *Ae. aegypti* populations during the Zika outbreak of 2017. Adult *Ae. aegypti* in Miami's Wynwood neighborhood were virtually eliminated by aerial spraying on two occasions, only to rebound to pre-spray levels in just three days [17]. In both studies, populations leveled off after rebounding. Release from larval competition does not fit the timing window for the rapid rebound observed following aerial spraying of adulticides, but temporary release from predation might allow such a rebound to occur. Off-target effects of insecticides on mosquito predators such as spiders, even non-lethal effects, might allow greater survival of adult mosquitos during the rebound period.

Mabel's orchard orbweaver spider (*Leucauge argyrobapta,* Tetragnathidae) is found in Florida, throughout the Caribbean, and from Mexico to Brazil, with congeners common throughout the North American continent [18]. In South Florida, *L. argyrobapta* is abundant in treed areas and around residences, constructing webs in vegetation, eaves of houses, porches, and patio furnishings. Both juvenile and adult *L. argyrobapta* are capable of subduing mosquitoes. *L. argyrobapta* produces viscous horizontal orb webs, and recycles sticky spiral capture threads nightly, as is typical of tetragnathid orb-weavers [12,19].

We explored the effects of sprayed permethrin, absorbed through web contact and web recycling, on the abilities of adult *L. argyrobapta* to repair their webs and capture live adult mosquitoes. Permethrin is a type I synthetic pyrethroid used routinely to control adult mosquitoes in Miami-Dade County (FL, USA) and elsewhere. Pyrethroids bind the pore of the voltage-gated sodium channel preventing its closure, thus causing the nerves to hyperpolarize and muscle action to cease [20].

2. Materials and Methods

We chose permethrin as our insecticide because of its wide use in mosquito control in Miami-Dade County. The treatment solution was a 0.0368% solution of permethrin in acetone (1000× dilution of Martin's 36.8% Permethrin SFR, Control Solutions Inc., Pasadena, TX, USA) with a cis/trans ratio 42–58%. Two negative controls were (1) the solvent carrier (AU582) diluted 1000× in acetone, and (2) acetone alone. By trial and error, we determined that this permethrin concentration did not kill any *L. argyrobapta* when applied to their webs. To put this 0.0368% concentration in context, one hour of foot contact with 2.1% permethrin impregnated paper is the LC-50 concentration for local *Ae. aegypti*, and 0.1% is the highest concentration that kills none of them (P.K. Stoddard, unpublished). Acetone is recommended as a pyrethroid carrier in arthropod toxicity studies because it evaporates quickly [21] and is tolerated well by arthropods [22]. Because of acetone's high volatility and our limitation on number of available and comparable spiders for allocation to treatment groups, we considered acetone-only a reasonable negative control and omitted an unsprayed control group. The perfect hunting success of spiders recorded on acetone-treated webs supported our choice. At just 0.06% of the solution, the amount of AU582 carrier solvent in the sprayed permethrin solution was insignificant and had no measurable effect on spider behavior.

L. argyrobapta were tested near a residence within a 0.4 ha plot of native hardwood forest in Miami-Dade County (25°30′00.25″, 80°28′03.48″). In March 2020, we removed 45 spiders from their webs, delicately so as not to damage the webs. We tore sections of the webs, both spiral mesh strands and several radii, to allow us to determine if the web had been recycled and reconstructed. We sprayed 15 webs with permethrin solution, 15 with acetone plus carrier, and 15 with acetone, dispensing of 2 mL of each solution per web using a hand sprayer. Webs receiving the different treatments were chosen at random with respect to the body size of the resident spider. We allowed webs to dry for 20 min then returned each spider to its original web. Webs were photographed 1 h after spraying and again 24 h later. We took two orthogonal measurements of web mesh diameter, both before spray application and again 24 h later, and noted whether the web had been substantially reconstructed (at least 75% replacement of spiral mesh).

In June 2020, we repeated the same web treatments, and determined the efficacy of treated webs and their resident spiders at retaining and subduing live mosquitoes, as well as whether the web had been substantially reconstructed. Ambient temperatures were higher in June than March, with overnight lows averaging 18 °C in March and 26 °C in June.

The mosquitoes we used were an even mix of wild *Ae. aegypti*, and *Wyeomyia vanduzeei*, captured locally with a BG-2 Sentinel trap (Biogents, Regensberg, Germany). Twenty-one webs were sprayed with permethrin and 16 with the acetone control. Using a mouth aspirator (John W. Hock Company, Gainesville, FL, USA), we aspirated five mosquitoes at random from the mosquito cage and propelled them toward an intact section of the web. Before conducting the trials, we practiced on untreated webs until we could reliably expel mosquitoes at the correct velocity to strike the web but not break the strands. Preliminary trials determined that undisturbed *L. argyrobapta* readily capture and consume any mosquito that sticks in the spiral web for at least three seconds (Figure 1). During the trials, we noted whether mosquitoes stuck to the web for 3+ s, whether the spider seized a stuck mosquito, and the latency for the spider to respond to the web strike.

Figure 1. (**A**) In the mosquito handling procedure, mosquitoes were propelled toward the web with a light puff of air, being careful not to blow air directly at the spider. (**B**) An adult female *L. argyrobapta* subdues a mosquito (*W. vanduzeei*) that had been propelled into the web.

Data on frequency of web repair were analyzed by comparing repaired and non-repaired webs as categorical variables using chi-square or Fisher's Exact Test to compare the treatment groups. We calculated web area from the two diameter measurements. Diameters of reconstructed webs were identical for every spider in the two control groups but differed in the permethrin-treated group, so web areas of spiders that attempted web reconstruction were compared before and after permethrin application using a paired T-test for unequal variance. Whether at least one mosquito adhering to each web was seized by spiders in the two treatment groups was evaluated with Fisher's Exact Test. Numbers of mosquitoes captured by the two treatment groups were compared using a 2-sample *t*-test.

3. Results

In March, all webs receiving the two control treatments had been substantially restored to their previous structure 24 h later, whereas none of the permethrin-treated webs were substantially restored ($p = 1.69 \times 10^{-10}$; $X^2 = 30$; 2 df). Six of the 15 permethrin-treated webs had been partially restored, and the rest were in the same torn condition as the day before. Of those webs that had been rebuilt, the mesh areas were, on average, just 42% of their original size ($p = 0.005$, t = 4.13, df = 6.8, paired T-test, unequal variance, 2-tailed). Visual inspection revealed that spacing of the adhesive spiral mesh strands was wider than on a normal web (Figure 2F). Most spiders on the permethrin-sprayed webs remained on the web, though several disappeared, having either abandoned the web or been predated.

In June, when we treated webs in preparation for mosquito capture, all control webs were substantially restored within 24 h, whereas just 29% (6 of 21) of the spiders on permethrin-treated webs had substantially restored their webs ($p < 0.0001$; Fisher's Exact Test). All permethrin-treated webs had been restored after three days, showing the interference at this dose was temporary.

The day after treating the webs in June, we tested the webs and spiders for mosquito-catching ability. Permethrin reduced the number of mosquitoes captured by 71%, independent of the reduced web area (Figure 3). Permethrin-treated webs were less effective at retaining the mosquitoes that struck them than the control webs; ($p = 0.00024$; t = 4.44, 15 df). Mosquitoes adhered to each of the 16 control webs for at least three seconds, and the resident spider seized a mosquito in each web. In the 21 permethrin-treated webs, we managed to stick a mosquito in what remained of 86% (18) of the webs; three webs could not retain a mosquito at all. Spiders on treated webs were less likely to seize a mosquito that did stick than spiders on control webs ($p < 0.0001$; Fisher's Exact test). In 67% of the webs that held a mosquito for at least 3 sec but that had not been substantially repaired (12 of 18), none of the spiders seized the mosquito. In the 21% of webs that were substantially reconstructed, each spider seized the mosquito immediately. Mosquitoes not seized by the spiders eventually wiggled free.

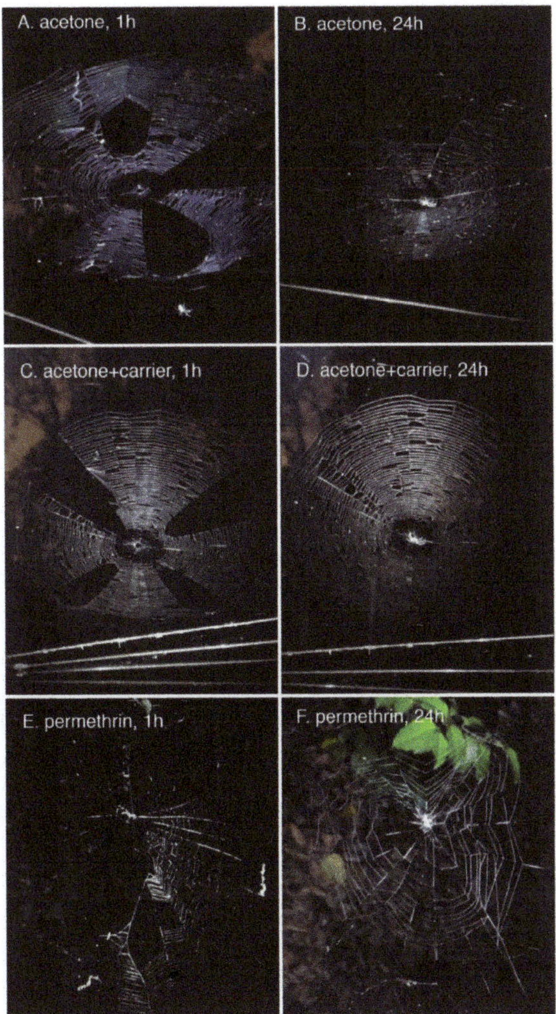

Figure 2. Side by side comparisons of spider webs before (**A**,**C**,**E**) and after (**B**,**D**,**F**) the residing orchard spiders were exposed to treatments. The control treatments (**A**–**D**) had no adverse effect on web reconstruction. For the first day after application, permethrin treatment (**E**,**F**) interfered with the resident spider's ability to reconstruct a normal web. The few webs that were partially reconstructed (6 of 15) had a significantly smaller web area, and irregular and wide spacing of adhesive spiral strands visible in the permethrin treatment after 24 h (**F**).

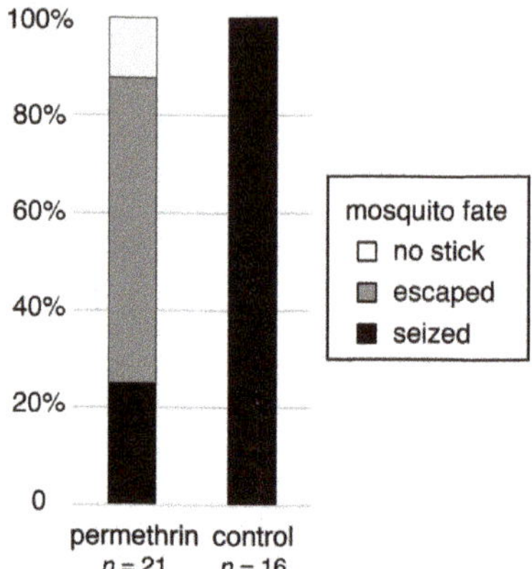

Figure 3. Spiders on permethrin-treated webs captured far fewer mosquitoes than those on control webs.

4. Discussion

Permethrin treatment of webs interfered with mosquito capture by spiders in at least two ways. After consuming the webs to recycle the silk, spiders were rendered sluggish. They were delayed in reconstructing the webs and produced small webs with irregularly spaced adhesive strands. Such webs were less likely to hold mosquitoes. Further, when mosquitoes did stick in these webs, the spiders most often did not react, allowing mosquitoes to struggle free. Similar effects have been documented for another synthetic pyrethroid, alpha-cypermethrin, which interferes with web reconstruction in the araneid orbweaver *Araneus diadematus* in the same ways we found for *L. argyrobapta* [23]. In *Tenuiphantes tenuis*, a common sheet-web spider of European cereal agricultural fields, the pyrethroid cypermethrin had similar detrimental effects on web construction and activity as well as reduced reproduction and shortened lifespan [24]. Spiders can still detect prey in webs with distorted structure [25], though they take about twice as long to do so, increasing likelihood of prey escaping. Reduced disc size would reduce capture efficiency proportionally.

Spiders exposed to permethrin in our study recovered in one to three days. A third of the spiders had recovered in one day in June, but none had in March. Seasonal temperature and humidity differences may have contributed to differences in speed of recovery. In South Florida, March is dry and cool, whereas June is hot and humid. Accordingly, diurnal mosquitoes are virtually absent in March, but their populations explode in June with the onset of the rainy season, so the June data better typify encounters between mosquitoes and spiders. Elevated temperature affects the spiders' response to pyrethroids in two opposing ways, hastening absorbance, which increases toxicity, but also increasing metabolism of the pesticide, which shortens recovery time for sublethal doses [26]. Pyrethroids also interfere with water transport in arthropods, leading to dehydration [26,27], which can affect all aspects of physiology.

The rapid rebound of mosquito populations over the three days following adulticide application [13,17] matches the three-day period of spider immobility found in this study and others [23,24,27]. Recovery of spiders' web construction and prey capture abilities would correspond to the return to the lower survival rate of adult *Ae. aegypti* three days after adulticide spray application. Non-lethal effects on spiders in this study have been seen in other species and with other pesticides [1,4,5,27]. Similarly, population increase of agricultural pests following reduction of spiders through pesticide application has been documented repeatedly [28]. Because spider immobility following pesticide application has significant consequences for control of agricultural pest infestation, the possibility of harm to spiders should be taken seriously in control of arbovirus epidemics as well. For that reason, research should be conducted to see whether nonlethal effects of pesticides on mosquito capture by spiders actually does enhance the speed of mosquito population rebound. Our study minimized environmental exposure by applying permethrin directly to the webs. A more direct comparison to typical mosquito control conditions would benefit from replicating these experiments using ULV spray applications of the pesticide of interest.

Mosquito populations are regulated by a mix of bottom-up and top-down processes including intraspecific and interspecific competition among larvae, and predation on all life stages [29–32]. Most studies of mosquito competition (bottom-up control) and predation (top-down control) focus on the larval stages. Small bats consume adults of mosquito species that fly in the open but not the urban mosquito species that stay close to dwellings and vegetation [33]. No evidence suggests bats eat enough mosquitoes to affect their populations [34]. Adult dragonflies can be seen hunting in the right places to catch mosquitoes in a variety of habitats, but, probably for logistic reasons, the literature is devoid of data on mosquito predation by adult dragonflies. Evidence for spiders as effective predators of adult mosquitoes is only slightly better. Web-building spiders are the dominant predators of flying insects in most terrestrial ecosystems [35], regulating prey populations, and, sometimes capturing far more prey in their webs than they can consume [36]. Density of pest insects is significantly lower in areas with higher densities of spiders [28]. Orb-weaving spiders, in particular, are key predators of dipterans including mosquitoes [37–39]. Accordingly, spider predation on mosquitoes is thought to have potential for control of dengue and malaria [40,41]. Our finding that spider predation of mosquitoes is temporarily compromised after pyrethroid exposure, combined with earlier findings that mosquito populations increase faster immediately following adulticide spraying [13,17] are consistent with the possibility that spiders play a significant role in suppression of urban/suburban mosquito populations. The argument thus far is correlative and would be illuminated by direct study of spider suppression of mosquitoes at the local population scale. Because spiders are so common in and around human dwellings, their potential as suppressors of adult mosquito populations warrants more direct experimental investigation.

Author Contributions: Conceptualization, S.N.R. and P.K.S.; methodology, S.N.R. and P.K.S.; software, P.K.S.; validation, S.N.R. and P.K.S.; formal analysis, P.K.S.; investigation, S.N.R. and P.K.S.; resources, P.K.S.; data curation, P.K.S.; writing—original draft preparation, S.N.R. and P.K.S.; writing—review and editing, P.K.S.; visualization, S.N.R. and P.K.S.; supervision, P.K.S.; project administration, P.K.S. All authors have read and agreed to the published version of the manuscript.

Funding: Funding for materials and supplies was provided by The Centers for Disease Control and Prevention (CDC), Southeastern Center of Excellence in Vector-borne Disease through the laboratory of Dr. Matthew DeGennaro at FIU.

Institutional Review Board Statement: The study was conducted according to the guidelines of the Declaration of Helsinki and authorized by the Institutional Animal Care and Use Committee of Florida International University (#IACUC-19-032, issued 22-Apr-2019).

Informed Consent Statement: Not applicable.

Data Availability Statement: Data are contained within the article.

Acknowledgments: Max Contador assisted in the web reconstruction experiment. The carrier solution was provided by Al Patterson at Control Solutions, Inc.

Conflicts of Interest: The authors declare no conflict of interest. The funders had no role in the design of the study; in the collection, analyses, or interpretation of data; in the writing of the manuscript, or in the decision to publish the results.

References

1. Richardson, J.R.; Fitsanakis, V.; Westerink, R.H.S.; Kanthasamy, A.G. Neurotoxicity of pesticides. *Acta Neuropathol.* **2019**, *138*, 343–362. [CrossRef] [PubMed]
2. Miliczky, E.R.; Calkins, C.O.; Horton, D.R. Spider abundance and diversity in apple orchards under three insect pest management programmes in Washington State, U.S.A. *Agric. For. Entomol.* **2000**, *2*, 203–215. [CrossRef]
3. Boyce, W.M.; Lawler, S.P.; Schultz, J.M.; McCauley, S.J.; Kimsey, L.S.; Niemela, M.K.; Nielsen, C.F.; Reisen, W.K. Nontarget effects of the mosquito adulticide pyrethrin applied aerially during a West Nile virus outbreak in an urban California environment. *J. Am. Mosq. Control Assoc.* **2007**, *23*, 335–339. [CrossRef]
4. Pekár, S.; Beneš, J. Aged pesticide residues are detrimental to agrobiont spiders (Araneae). *J. Appl. Entomol.* **2008**, *132*, 614–622. [CrossRef]
5. Pasquet, A.; Tupinier, N.; Mazzia, C.; Capowiez, Y. Exposure to spinosad affects orb-web spider (*Agalenatea redii*) survival, web construction and prey capture under laboratory conditions. *J. Pest. Sci.* **2016**, *89*, 507–515. [CrossRef]
6. Bogya, S.; Marko, V.; Szinetar, C. Effect of pest management systems on foliage- and grass-dwelling spider communities in an apple orchard in Hungary. *Int. J. Pest. Manag.* **2000**, *46*, 241–250. [CrossRef]
7. Samu, F.; Matthews, G.A.; Lake, D.; Vollrath, F. Spider webs are efficient collectors of agrochemical spray. *Pestic. Sci.* **1992**, *36*, 47–51. [CrossRef]
8. Townley, M.A.; Tillinghast, E.K. Orb web recycling in *Araneus cavaticus* (Araneae, Araneidae) with an emphasis on the adhesive spiral component, gabamide. *J. Arachnol.* **1988**, *16*, 303–319.
9. Carico, J.E. Web removal patterns in orb-weaving spiders. In *Spiders: Webs, Behavior, and Evolution*; Shear, W.A., Ed.; Stanford University Press: Stanford, CA, USA, 1986; pp. 306–318.
10. Breed, A.L.; Levine, V.D.; Peakall, D.B.; Witt, P.N. The fate of the intact orb web of the spider *Araneus diadematus* Cl. *Behaviour* **1964**, *23*, 43–60.
11. Peakall, D.B. Conservation of web proteins in the spider, *Araneus diadematus*. *J. Exp. Zool.* **1971**, *176*, 257–264. [CrossRef]
12. Opell, B.D. Economics of spider orb-webs: The benefits of producing adhesive capture thread and of recycling silk. *Funct. Ecol.* **1998**, *12*, 613–624. [CrossRef]
13. Focks, D.A.; Kloter, K.O.; Carmichael, G.T. The impact of sequential ultra-low volume ground aerosol applications of malathion on the population-dynamics of *Aedes aegypti* (L). *Am. J. Trop. Med. Hyg.* **1987**, *36*, 639–647. [CrossRef] [PubMed]
14. Alto, B.W.; Lampman, R.L.; Kesavaraju, B.; Muturi, E.J. Pesticide-induced release from competition among competing *Aedes aegypti* and *Aedes albopictus* (Diptera: Culicidae). *J. Med. Entomol.* **2013**, *50*, 1240–1249. [CrossRef] [PubMed]
15. Muturi, E.J.; Costanzo, K.; Kesavaraju, B.; Lampman, R.; Alto, B.W. Interaction of a pesticide and larval competition on life history traits of *Culex pipiens*. *Acta Trop.* **2010**, *116*, 141–146. [CrossRef]
16. Kesavaraju, B.; Brey, C.W.; Farajollahi, A.; Evans, H.L.; Gaugler, R. Effect of malathion on larval competition between *Aedes albopictus* and *Aedes atropalpus* (Diptera: Culicidae). *J. Med. Entomol.* **2011**, *48*, 479–484. [CrossRef]
17. Stoddard, P.K. Managing *Aedes aegypti* populations in the first Zika transmission zones in the continental United States. *Acta Trop.* **2018**, *187*, 108–118. [CrossRef]
18. Ballesteros, J.A.; Hormiga, G. Species delimitation of the North American orchard-spider *Leucauge venusta* (Walckenaer, 1841) (Araneae, Tetragnathidae). *Mol. Phylogenet. Evol.* **2018**, *121*, 183–197. [CrossRef]
19. Opell, B.D.; Bond, J.E.; Warner, D.A. The effects of capture spiral composition and orb-web orientation on prey interception. *Zoology* **2006**, *109*, 339–345. [CrossRef]
20. Dong, K.; Du, Y.; Rinkevich, F.; Nomura, Y.; Xu, P.; Wang, L.; Silver, K.; Zhorov, B.S. Molecular biology of insect sodium channels and pyrethroid resistance. *Insect Biochem. Mol. Biol.* **2014**, *50*, 1–17. [CrossRef]
21. World Health Organization. Guidelines for Efficacy Testing of Insecticides for Indoor and Outdoor Ground Applied Space Spray Applications. 2019. Available online: http://apps.who.int/iris/bitstream/handle/10665/70070/WHO_HTM_NTD_WHOPES_2009.2_eng.pdf?sequence=1 (accessed on 1 February 2021).
22. Hoskins, W.M.; Gordon, H.T. Arthropod resistance to chemicals. *Annu. Rev. Entomol.* **1956**, *1956*, 89–122. [CrossRef]
23. Samu, F.; Vollrath, F. Spider orb web as bioassay for pesticide side-effects. *Entomol. Exp. Appl.* **1992**, *62*, 117–124. [CrossRef]
24. Shaw, E.M.; Wheater, C.P.; Langan, A.M. The effects of cypermethrin on *Tenuiphantes tenuis* (Blackwall, 1852): Development of a technique for assessing the impact of pesticides on web building in spiders (Araneae: Linyphiidae). *Acta Zool. Bulg.* **2006**, *1*, 173–179.
25. Mulder, T.; Mortimer, B.; Vollrath, F. Functional flexibility in a spider's orb web. *J. Exp. Biol.* **2020**, *223*, jeb234070. [CrossRef] [PubMed]
26. Everts, J.W.; Willemsen, I.; Stulp, M.; Simons, L.; Aukema, B.; Kammenga, J. The toxic effect of deltamethrin on linyphiid and erigonid spiders in connection with ambient-temperature, humidity, and predation. *Arch. Environ. Contam. Toxicol.* **1991**, *20*, 20–24. [CrossRef]
27. Pekár, S. Side effect of synthetic pesticides on spiders. In *Spider Ecophysiology*; Nentwig, W., Ed.; Springer: Berlin, Germany, 2012; pp. 415–427.
28. Marc, P.; Canard, A.; Ysnel, F. Spiders (Araneae) useful for pest limitation and bioindication. *Ag. Ecosyst. Environ.* **1999**, *74*, 229–273. [CrossRef]

29. Juliano, S.A. Species interactions among larval mosquitoes: Context dependence across habitat gradients. *Annu. Rev. Entomol.* **2009**, *54*, 37–56. [CrossRef]
30. Gilpin, M.E.; Mcclelland, G.A.H. Systems-analysis of the yellow-fever mosquito *Aedes aegypti*. *Forts. Zool.* **1979**, *25*, 355–388.
31. Dye, C. Intraspecific competition amongst larval *Aedes aegypti*—Food exploitation or chemical interference. *Ecol. Entomol.* **1982**, *7*, 39–46. [CrossRef]
32. Dye, C. Competition amongst larval *Aedes aegypti*—The role of interference. *Ecol. Entomol.* **1984**, *9*, 355–357. [CrossRef]
33. Wray, A.K.; Jusino, M.A.; Banik, M.T.; Palmer, J.M.; Kaarakka, H.; White, J.P.; Lindner, D.L.; Gratton, C.; Peery, M.Z. Incidence and taxonomic richness of mosquitoes in the diets of little brown and big brown bats. *J. Mammal.* **2018**, *99*, 668–674. [CrossRef]
34. Wetzler, G.C.; Boyles, J.G. The energetics of mosquito feeding by insectivorous bats. *Can. J. Zool.* **2017**, *96*, 373–377. [CrossRef]
35. Nyffeler, M.; Benz, G. Spiders in natural pest-control—A review. *J. Appl. Entomol.* **1987**, *103*, 321–339. [CrossRef]
36. Michalko, R.; Pekar, S.; Entling, M.H. An updated perspective on spiders as generalist predators in biological control. *Oecologia* **2019**, *189*, 21–36. [CrossRef] [PubMed]
37. Nyffeler, M.; Benz, G. Foraging ecology and predatory importance of a guild of orb-weaving spiders in a grassland habitat. *J. Appl. Entomol.* **1989**, *107*, 166–184. [CrossRef]
38. Henaut, Y.; Garcia-Ballinas, J.A.; Alauzet, C. Variations in web construction in *Leucauge venusta* (Araneae, Tetragnathidae). *J. Arachnol.* **2006**, *34*, 234–240. [CrossRef]
39. Muma, M.H. Spiders in Florida citrus groves. *Fla. Entomol.* **1975**, *58*, 83–90. [CrossRef]
40. Strickman, D.; Sithiprasasna, R.; Southard, D. Bionomics of the spider, *Crossopriza lyoni* (Araneae, Pholcidae), a predator of dengue vectors in Thailand. *J. Arachnol.* **1997**, *25*, 194–201.
41. Ndava, K.; Diaz Llera, S.; Manyanga, P. The future of mosquito control: The role of spiders as biological control agents: A review. *Int. J. Mosq. Res.* **2018**, *5*, 6–11.

Article

Reduction of Pesticide Use in Fresh-Cut Salad Production through Artificial Intelligence

Davide Facchinetti [1,*], Stefano Santoro [1,*], Lavinia Eleonora Galli [1], Giulio Fontana [2], Lorenzo Fedeli [3], Simone Parisi [3], Luigi Bono Bonacchi [3], Stefan Šušnjar [3], Fabio Salvai [4], Gabriele Coppola [4], Matteo Matteucci [2] and Domenico Pessina [1]

1 Department of Agricultural and Environmental Sciences—Production, Territory, Agroenergy, University of Milan, 20133 Milan, Italy; lavinia.galli@unimi.it (L.E.G.); domenico.pessina@unimi.it (D.P.)
2 Department of Electronics, Information and Bioengineering, Politecnico di Milano, 20133 Milan, Italy; giulio.fontana@polimi.it (G.F.); matteo.matteucci@polimi.it (M.M.)
3 Alta Scuola Politecnica, Politecnico di Milano, 20133 Milan, Italy; lorenzo1.fedeli@mail.polimi.it (L.F.); simone1.parisi@mail.polimi.it (S.P.); luigibono.bonacchi@mail.polimi.it (L.B.B.); stefan.susnjar@asp-poli.it (S.Š.)
4 Alta Scuola Politecnica, Politecnico di Torino, 10129 Torino, Italy; fabio.salvai@asp-poli.it (F.S.); gabriele.coppola@asp-poli.it (G.C.)
* Correspondence: davide.facchinetti@unimi.it (D.F.); stefano.santoro@unimi.it (S.S.)

Citation: Facchinetti, D.; Santoro, S.; Galli, L.E.; Fontana, G.; Fedeli, L.; Parisi, S.; Bonacchi, L.B.; Šušnjar, S.; Salvai, F.; Coppola, G.; et al. Reduction of Pesticide Use in Fresh-Cut Salad Production through Artificial Intelligence. *Appl. Sci.* **2021**, *11*, 1992. https://doi.org/10.3390/app11051992

Academic Editor: Giuseppe Manetto

Received: 31 January 2021
Accepted: 19 February 2021
Published: 24 February 2021

Abstract: Incorrect pesticide use in plant protection often involve a risk to the health of operators and consumers and can have negative impacts on the environment and the crops. The application of artificial intelligence techniques can help the reduction of the volume sprayed, decreasing these impacts. In Italy, the production of ready-to-eat salad in greenhouses requires usually from 8 to 12 treatments per year. Moreover, inappropriate sprayers are frequently used, being originally designed for open-field operations. To solve this problem, a small vehicle suitable for moving over rough ground (named "rover"), was designed, able to carry out treatments based on a single row pass in the greenhouse, devoted to reduce significantly the sprayed product amount. To ascertain its potential, the prototype has been tested at two growth stages of some salad cultivars, adopting different nozzles and boom settings. Parameters such as boom height, nozzle spacing and inclination, pump pressure and rover traveling speed were studied. To assess the effectiveness of the spraying coverage, for each run several water-sensitive papers were placed throughout the vegetation. Compared to the commonly distributed mixture volume (1000 L/ha), the prototype is able to reduce up to 55% of product sprayed, but still assure an excellent crop coverage.

Keywords: greenhouse; precision spraying; rover; environment protection; operator's health; consumer's safety

1. Introduction

Due to the progressive increase of the world population, which, according to the Food and Agriculture Organization (FAO), will exceed 9 billion inhabitants in 2050 [1], the agricultural sector will have to satisfy an increasing demand for food. To ensure high productions, a crucial role will be played by pesticides.

According to FAO, pesticides include any substance, or mixture of substances of chemical (natural or synthetic) or biological ingredients for repelling, destroying, or controlling any pest, or regulating plant growth [2]. Indeed, they can help to protect seeds and safeguard crops from unwanted plants, insects, bacteria, fungi and rodents. Pesticides include a wide range of herbicides, insecticides, fungicides, rodenticides, and nematicides.

Usually, pesticide is a more general term that comprises plant protection products (PPPs) which are aimed to protect crops or desirable/useful plants. PPPs contain at least one active substance and have different functions such as to protect plants or plant products against pests or diseases before or after harvest, to influence the life processes of the plant, to preserve plants products while destroying or to prevent the growth of undesired plants or parts of plants [3]. When pesticides are used irresponsibly, they might involve health risks, can have negative environmental impacts, on the soil and water, can reduce biodiversity, and, in some cases, decrease crop yield.

The total pesticides used in agriculture remained stable in 2018 compared to 2017, with a light decrease from 4.15 Mt to 4.12 Mt, but in the last 30 years, from 1990 to 2018, the global use of pesticides in agriculture has increased from 1.80 to 2.66 kg/ha. The global application of pesticides increased in this period both for herbicides, fungicides, and insecticides [4].

1.1. Pesticide Use in Europe

Europe increased pesticide use in agriculture in the 2010s compared to the 1990s by just 5%, most likely due to the stringent European Common Agricultural Policy that came into force, which monitors and controls the use of pesticides. Indeed, due to the application of the Directive 2009/128/CE of the European Commission, member states adopted national action plans to set up their quantitative objectives, targets, measures, and timetables to reduce risks and impacts of pesticides use on human health and environment, to encourage the development and introduction of integrated pest management and alternative approaches or techniques to reduce dependency on the use of pesticides [5]. Pesticide use per area of cropland was approx. 1.66 kg/ha in 2018 [4]; the average use of pesticides from 1990 to 2018 has been 465,556 tons [6]. Four EU member states, Germany, Spain, France and Italy, represent by themselves over two-thirds of the total EU pesticide sales volume. These countries are also the main agricultural producers in the EU, with collectively 51% of the total EU utilized agricultural area (UAA), and 49% of the total EU arable land. In terms of the categories of pesticides sold, the highest sales volume in 2018 was for fungicides and bactericides (45%), followed by herbicides (32%) and insecticides and acaricides (11%) [7].

1.2. Pesticide Use in Italy

Italian agriculture has one of the highest uses of PPPs in Europe. In 2016, according to the Italian National Institute of Statistics (ISTAT) [8], the use was 7.22 kg/ha of active substances, corresponding to approximately 124,000 t per year. The value of the Italian market for crop protection products has increased by 43.6% in the last 10 years. The reason for this variation was the constant improvement of the mix of products that, due to a lower dose rate, has led to an increase in unit price. A strong decrease in the quantities used has been observed (−22%), shifting from 141,200 to 109,860 t from 1990 to 2015. In terms of active substances, the categories most concerned by the introduction of innovative molecules with a low dosage are mainly represented by fungicides and herbicides, which have determined the consistent decrease. A survey, carried out by the European Food Safety Authority (EFSA) in 2016, has highlighted the high qualitative standards of the Italian products, thanks to a control system extremely stringent and efficient that ensure a high safety level to consumers. Only 1.2% of the sample analyzed proved to be irregular, compared to a European average of roughly 2.9% [9].

1.3. Fresh-Cut Products

According to the International Fresh-Cut Produce Association (IFPA), fresh-cut products are fruits or vegetables that have been trimmed and/or peeled and/or cut into 100% usable products. Fresh-cut products are cut, washed, packaged, and maintained with refrigeration. They are in a raw state and even though minimally processed, they remain in a fresh state, ready to eat or cook. Since their origin in Europe in the early 1980s, they have become more and more common in consumer baskets. The innovation represented by this sector involves the technologies adopted in growing, processing and marketing. Fresh-cut products such as ready-to-eat salads require substantial capital investment in plants and machinery [10]. This sector is characterized by intensive cultivation that requires large use of chemical products, such as fertilizers and pesticides.

1.4. Fresh-Cut Salad Sector in Italy

The agricultural area devoted to the production of fresh-cut vegetables in Italy is about 6500 ha. Production takes place mainly in plastic tunnels (in northern Italy), which are called "Bergamasca", or multitunnels (in southern Italy) [11]. The most common structure adopted is plastic tunnels of semicircular shape, with a width of 8–10 m, a length of 50–100 m and a height of 4 m. Usually, the greenhouse is divided into four rows of crops, with a typical inter-row of 1.5 m, separated by small ruts of about 0.2 m of width, being the paths for the tractor wheels.

In 2015, Italy produced 110,000 t of fresh-cut vegetables, with a value of € 744 million. Of the total value, lettuce comprised 75.4%, followed by wild rocket (9.5%), spinach (4.5%), and Swiss chard (1.3%) [10].

Normally, the Italian ready-to-eat salad producers cultivate various types of salads, such as lettuce (*Lactuca sativa* L.), wild rocket (*Diplotaxis tenuifolia* L.), spinach (*Spinacia oleracea* L.), lamb's lettuce (*Valerianella olitoria* L.), some subspecies of chicory (*Cichorium intybus* L. subspecies), and many others. These salads are grown in five to six cycles per year in the same greenhouse, with a duration of the production cycle that may vary from 20 to 90 days, according to the season. These plants grow under high crop density, with a lack of adequate crop rotation, and require a high number of pesticides treatments to avoid severe product losses.

Pesticide treatments are typically performed two times during each life cycle of the plants; therefore, the number of treatments is estimated at 10 to 12 every 12 months, and the quantity of volume sprayed currently used is 1000 L/ha. For the treatment, machines designed for open fields are typically used, instead of machine optimized for greenhouses. These are constituted of a tank (usually from 300 to 1100 L capacity) where the pesticide mixed in water is contained, a hydraulic circuit which distributes the liquid up to the set of nozzles in charge of the spraying, and, sometimes, a fan that uses air to move the vegetation and partially avoid the drift effect of the droplets (movement of spray particles away from the target area). Usually, the spraying is performed through a simple boom, with no particular setup parameters, attached to the tank, and driven by a small tractor.

Plant protection products are sold in liquid form or in soluble powder; to be used, they must be carefully mixed with water in the doses reported on the product label. Doses describe the amount of product per quantity of water (g/L) and the recommended quantity of PPPs per hectare (kg/ha); the last item of data, which is redundant, shows separately the water amount to be added in the mixture per unit area (L/ha). Regulators require that the quantity prescribed on the product label be used, so farmers cannot define an optimal amount of spray in different ways, as it is legally forbidden in Italy.

It has been established that for the pesticides acting by contact, the correct dose to be used should not be referred to the field area, but rather to the overall surface of the leaves. However, during the growing period, the overall surface of a crop changes remarkably. Taking into account these dynamics, the scope of this work is to develop an artificial vision system supported by artificial intelligence, able to define optimal levels of mixture (PPP + water) related to different growth stages, to assure the highest protection effect, minimizing waste at the same time.

2. Materials and Methods

In greenhouses, the salad is picked approximately after one or two months from when it is planted, so five to six cycles per year are possible, resulting in 10 to 12 pesticide treatments every twelve months. This number of treatments is excessive, and it can create some problems regarding the environmental impact. Specifically, the spraying modes could be significantly improved, since no dedicated machines for greenhouses have been developed yet and decisions about when and what to treat are often based on personal experience. This leads often to an overestimation of the PPPs amount to be sprayed, to guarantee that also the lower layer of leaves is reached.

In this scenario, a structured solution to tackle the main criticalities of treatments in the fresh-cut salad cultivation is proposed: a real-time advanced monitoring machine, which aims at improving the treatment by deploying robotics and artificial intelligence (AI) techniques to define how to properly spray, given the specific characteristics of the plant. At the base of this principle, we have an autonomous rover adapting the controllable characteristics of the treatment (e.g., the pressure of spraying, volume of air used, or height of nozzles from the ground) was designed and built. The rover was equipped with an advanced boom, according to what is detected in real-time from sensors, about the current state of the plant. The rover is completely autonomous and, ideally, it does not require any human assistance, saving the cost of labor or at least significantly reducing it.

As for the motion, the most effective solution is to adopt a small size rover, ideally performing the treatment on one row at a time. In this way, the prototype design is simplified, and the manufacturing costs are minimized. The rover is coupled to an advanced boom, to guarantee large control possibilities, leading to more improvements in treatment efficiency. The advanced boom can control the treatment by adjusting its height, the air distribution flow, the nozzle distance and angle, the pressure control and other operation parameters.

2.1. Main Subsystems of the Machine

The main subsystems of the designed machine are: data gathering, spraying circuit, boom structure and actuation, rover structure and motion, and navigation computing/electronics. The aim of the data gathering subsystem is to acquire information regarding the plant and the treatment, carried out thanks to a 3D camera, a pressure sensor and a flow sensor. The 3D camera is by far the most important component, both in terms of information output and cost; its core function is to determine the type and growth stage of the plant, while allowing a visual odometer to measure the speed of the rover, a fundamental parameter for the treatment. The water pressure and flow sensors instead allow a closed-loop control of the droplet size and product amount. While in theory the pressure-flow curve of the nozzles is known, and therefore only one of these sensors would be needed, the presence of both allows to detect faults (clogged nozzles, disconnected pipes, air in the pump, etc.) and consequently stop the treatment, thus allowing the user to solve the problem immediately.

The spraying circuit is the subsystem with the more traditional architecture. The water line starts from the plastic tank and, through a high-pressure (up to 8 bar) variable speed membrane pump, delivers the water to the nozzles. To allow different spacing, two rows of nozzles were installed, driven by an electrovalve to allow using only one row, so resulting in a 50 cm distance between active nozzles, or both, reducing this distance to 25 cm. The 50 cm distance was considered because it is the standard generally adopted on sprayer booms, while the addition of the second row allows multiangle treatment and lower dispersion of the droplets when reducing the boom height from the ground. A low-pressure (0.2 bar) agitation pump recirculates the water in the tank, to avoid sedimentation when a powder-based of PPP is used. The air line is composed only of a variable speed fan and a flexible plastic air sleeve to direct the flow over the treatment area.

The boom structure is an aluminum body-on-frame. The selection of aluminum as the main material, despite the higher cost, is due to the necessity to keep this component as light as possible since the overhang outside the track and the height from the ground have to be adjustable. To change the height, within a range of 30 cm, a linear electric actuator was used, to cover the optimal settings for both the defined nozzle distances.

The rover structure is a mild steel body-on-frame, coated with high-temperature powder paint to increase durability. Due to the low working speed, no suspension system was implemented. The rear wheels have independent in-hub motors, while the front wheels are free-pivoting. This configuration allows controlling the motion without a steering system, by imposing the speed of the right and left wheel. The motors are driven from two independent power drives that implement a closed-loop speed control and are directly connected to the battery. The battery is a 48 V Li-ion pack with a waterproof casing. For a suitable safety of the operations, an emergency switch is put on the battery for the disconnection.

The navigation system is based on a differential real-time kinematic GPS. This system consists of three sensors and antennas, two on the rover and one in a fixed reference position on the ground. Thanks to the real-time kinematic algorithm the relative position of the antennas can be known with precision in the subcentimetric range. The two antennas on the rover allows to determine differentially its spatial orientation, with a typical uncertainty of 0.2°. The data coming from the GPS are fused with those obtained from the inertial measurement unit (IMU) and the 3D camera, to allow excellent accuracy and to stop safely the system in case of malfunctions. The data taken from all the navigation sensors, as well as the data gathering subsystem, are then collected and elaborated from the main computing unit, a Linux-based x86 Personal Computer (PC), to determine the movement and treatment actuator response. The PC is paired with a microcontroller and a power unit to electrically interface with the sensors and actuators. A separate power supply is used for the PC, while the microcontroller board is powered directly through the PC. To evaluate the potential of this solution, a prototype has been designed and built to run on-field tests to define the optimal way of spraying, given the characteristics of the plants.

2.2. *Prototype*

To verify the potential of the prototype, an advanced boom equipped with the camera was realized before the complete rover, to test the sprayer while varying all the considered parameters, and therefore to validate the reduction of the pesticide usage. In this paper, the detailed design of the advanced boom and the results of its validation are reported, before any industrialization step. Focusing on advanced boom validation instead of the full system, allowed to manage properly the basic functionalities. In particular, the prototype detailed in this paper includes:

- A simplified structure made with wooden panels, while the actuation of the boom was eliminated, and the "legs" of the prototype were used to manually change the height between tests.
- A small tank, of 20 L capacity, however sufficient for carrying out several test sets.
- No flow sensors: new nozzles are used, the nominal flow rate is therefore known and so no pressure sensor is needed to obtain all the flow parameters.
- No navigation hardware. The testing will be carried out in a straight line.
- Laptops instead of a dedicated embedded PC.
- A main power supply instead of a battery. Moreover, due to the reduced power, the main circuit voltage was reduced from 48 V to 12 V.
- A geared motor instead of the in-hub motors. The gear was used to drive a chain with one extremity fixed to the wall and so regulate the test bench speed.
- Two cheap pumps have been used instead of a professional model. The quality of the pump is not relevant to the test results since the flow conditions are directly imposed.

The CAD model (Figure 1) shows the design and the overall structure of the test bench manufactured for quantitative evaluation of the proposed solution.

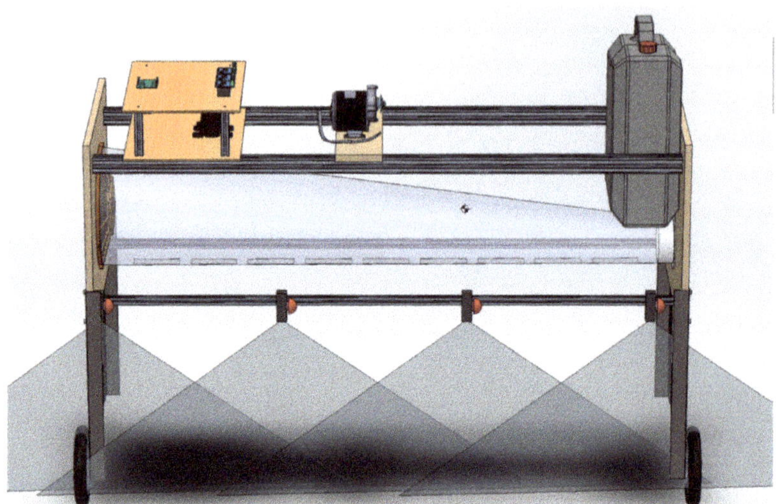

Figure 1. CAD model of the prototype.

2.3. Field Tests

The main goal was to test the prototype in different working conditions considering various operating parameters, to find the most suitable spraying setting in given crop conditions. Two test sessions were carried out, as the pesticides are distributed at different stages of the plants' growth. The height and the width of the salad leaves were taken into account as representative of the crop. In detail, the parameters investigated were:

- the nozzles' typology: the nozzle type affects the drops' dimension;
- nozzle distance: the distance affects the area covered by the drops, and in particular the spraying overlap between the adjacent nozzles, in order to obtain the maximum distribution uniformity;
- nozzle height: the height of the nozzles from the top of the crop affects the size of the drop-covered area;
- rover speed: the speed affects the amount of mixture distributed on a given surface;
- pressure: the variation of the spraying pressure changes the drop size, affecting the coverage of the crop, but also the possible drift of the mixture.

A total of 37 tests were carried out, arranging different combinations of the shown features. To check the amount and evaluate the spraying features, 75 water-sensitive papers were used for each test; the sensitive papers were placed in the crop in predefined ways, and after each run collected and analyzed. A depth camera was used to record each test and an image processing algorithm has been implemented to identify the plants and measure height and covered area.

For the tests, radicchio (*Cichorium intybus* L.) was selected as the main crop, being representative of most used salads in the ready-to-eat salad sector.

2.3.1. Real-Time Measurement of the Plants' Features

The camera used for the tests is the Intel RealSense™ Depth Camera D435i [12]. Apart the relatively low cost (about 200 $), this model allows at the same time to measure distances and to collect images; moreover, this specific model is equipped with an integrated IMU sensor. The algorithm for image processing has been implemented in C++ via the RealSense library [13], as it contains all the functions needed to use the camera.

The algorithm structure for image processing implements a classical pipeline. At first, some filters [14] are defined and applied to the data acquired by the camera, to reduce the noise and improve image quality, as follows:

- decimation filter: the resolution of the depth scene is reduced to reduce the necessary computation power;
- spatial filter: an edge-preserving filter with a performance that can be used in real-time;
- temporal filter: a filter that uses the information of the previous frames to adjust the current frame. This is a useful function, above all in static conditions.

After data acquisition and filtering, color detection and segmentation algorithms were used to find the plants. The OpenCV library [15] has been used to implement a simple algorithm [16] assuming, in a greenhouse scenario, the most common salad color is in the range of green, while the terrain is typically earth brown. Based on this simple assumption, it was necessary to find the right range of greens to detect the plants; although a possible solution could have been to work in the red, green, and blue (RGB) domain [17], since the frames were already in this format, the color values of this domain are extremely sensitive to light intensity, and therefore the best option was to convert each frame into the hue saturation value (HSV) domain [18]. Once the RGB frame is converted into HSV, a binary mask is computed to find where a specified color is.

As the tests were not conducted in a greenhouse, the assumption on the contrast between the salads' and the terrain's colors was not valid, and we had to tune the range also considering the background of the images, containing grey (the floor color). Due to some reflection, this grey area in the HSV analysis presented some "yellow" components, very close to the "green" part of the HSV spectrum; the choice of a proper threshold turned in to exclude some parts of the salads not to detect erroneously the ground. For real applications in greenhouse this should not happen, because of the clearer distinction between the color of the terrain and the salads.

Knowing where the plants are, it is possible to extract useful information like the height of the plants, from the depth component of the camera, and the percentage of the covered area. To work in real-time, some additional information on the plants is needed; indeed, to reduce the computational power of the algorithm and increase the throughput, the analysis is performed on a smaller area of the frame chosen in the middle of the frame, with a height which is one eighth of the total height of the image and a width that is two-thirds of the frame width. In Figure 2, the red rectangle identifies the plants detected, while the blue one indicates the area analyzed. Assuming the plants we want to analyze are in the middle of the frame, the algorithm measures the distance of the floor/terrain using a strip to the left and a strip to the right of the area under analysis. In doing this, all the "green" pixels are discarded, while the other measures are stored in a vector. The same process is done also for the plants: the algorithm considers all the pixels that are in the intersection of the area under analysis (the blue rectangle) and where the plants are identified; if the pixel is green, the distance is stored in another vector.

Having two vectors, respectively, for the distance measurements of the floor and the distance measurements of the plants, we compute the floor and plant distances using the median of the available samples as the mean would have been too sensible to the presence of several outliers. In a greenhouse scenario, the availability of these two values is sufficient to compute the median height of the plants in the area taken into account for the analysis. During the laboratory tests, the plants were placed in some trays, thus, the algorithm measures the height of the plants as:

$$\text{median\_plant\_height} = \text{median\_floor\_distance} - \text{median\_plants\_distance} - \text{trays\_height}$$
$$(\text{trays\_height} = 0.06\ \text{m}).$$

The vision algorithm measures also the percentage of the analyzed area covered by the plants counting the number of green pixels recognized as plants and computes the ratio between this number and the total number of pixels inside the area. This last method uses 2D information, thus it does not extract volumetric characteristics of the plants.

Figure 2. Image processing algorithm output.

2.3.2. Evaluation of the Spraying Quality and the Amount of Spray Liquid (Water) Used

Seventy-five square-shaped water-sensitive papers of about 1 cm^2 area were placed on the upper surface of leaves (25 samples), on the lower surface (25 samples), and on the ground (25 samples). Salads were planted in eight bedding plant trays, each containing 6×8 smaller trays, each of them containing a plant. The dry water-sensitive papers were gently stuck to the plant leaf using a natural glue, and these remained there after the end of the treatment (i.e., the passage of rover with spraying water on the eight trays), including the time needed for the plants and papers to become dry again.

To perform a more systematic analysis and obtain quantitative results, the samples were classified and then attached to a blank A4 paper for image processing. A dedicated algorithm was used to extract every single water-sensitive paper and to perform the color analysis.

Dry water-sensitive papers are typically yellow, but when completely wet become blue. The parts of papers not well sprayed become light blue. The image processing algorithm is able to detect pixels with sufficient color saturation, meaning that they are close to the blue threshold value, defined from the sample of scanned papers. Since the samples were not perfectly squared and their area was not exactly 1 cm^2, and during gluing them on white A4 paper they could not be perfectly aligned, the image to be processed also needed detection of all other nonwhite colors (Figure 3). This way, a few parameters were introduced, defined for every single extracted water-sensitive sample:

- N_{total} is the number of nonwhite pixels, i.e., the number of pixels in a 2D image representing the water-sensitive paper.
- N_{blue} is the number of blue pixels, i.e., those pixels in the 2D image of the water-sensitive paper that represent parts of the paper covered by water.
- $x_{c,total}$ is the x coordinate of the geometric center of all nonwhite pixels in the 2D image.
- $y_{c,total}$ is the y coordinate of the geometric center of all nonwhite pixels in the 2D image.
- $x_{c,blue}$ is the x coordinate of the geometric center of all blue pixels in the 2D image.
- $y_{c,blue}$ is the y coordinate of the geometric center of all blue pixels in the 2D image.
- $c = \frac{N_{blue}}{N_{total}}$ is the coverage, can be expressed in percentage as $\frac{N_{blue}}{N_{total}} \times 100\%$ indicates what percentage of the area of the water-sensitive paper is covered by water or, in more practical terms, sprayed well.

- $d = \sqrt{(X_{c,blue} - X_{c,total})^2 + (Y_{c,blue} - Y_{c,total})^2}$ is the distance (offset) between the two geometric centers defined above; the lower the value, the more centered spraying is performed.

Figure 3. Some examples of single water-sensitive papers extracted by extraction algorithm (on the left in each pair) and black and white images obtained after transforming all 'blue pixels' to 'black' (on the right in each pair).

The final output of the water-sensitive papers analysis algorithm is a report containing the following parameters for each test and each region (specific position—upper leaf, lower leaf, ground):

- $N_{insufficient}$ is the number of insufficiently sprayed samples, i.e., those papers for which the coverage c is below the adopted threshold $t = 7\%$. The threshold value was adopted experimentally, after running the program for different values and comparing with the exemplary papers from reference [19].
- $N_{sufficient}$ is the number of all other samples from the same test and region; in total $N = N_{sufficient} + N_{insufficient} \leq 25$ is the number of papers in a single test, from a single specific region.
- $\bar{c}$ is the average coverage for the single test, defined as $\bar{c} = \frac{\sum_{i=1}^{N} c_i}{N}$, c_i is the coverage of every single water-sensitive paper belonging to this test and region.
- $\sigma_c = \sqrt{\frac{\sum_{i=1}^{N} (c_i - \bar{c})^2}{N}}$ is the standard deviation of coverage for the given test and region.
- $\bar{d}$ is the average distance (offset) for the single test, defined as: $\bar{d} = \frac{\sum_{i=1}^{N} d_i}{N}$, where d_i is the offset for every single water-sensitive sample belonging to the considered test and region.
- $\sigma_d = \sqrt{\frac{\sum_{i=1}^{N} (d_i - \bar{d})^2}{N}}$ is the standard deviation of distance (offset) for the given test and region.

Another important aspect of the analysis is the amount of mixture (water) distributed. It can be estimated, knowing the flow rate of each nozzle, the speed at which the rover moves and, to refer to the area covered, the distance between the nozzles. Such quantity, called "coverage", is defined as

$$C = \left(\frac{l}{ha}\right) = \frac{Q}{v\Delta s} \times 10^4 \quad (1)$$

where:

$Q\left(\frac{l}{s}\right)$ is the nozzle flow rate

$v\left(\frac{m}{s}\right)$ is the travelling speed of the rover

$\Delta s(m)$ is the distance between the nozzles.

To calculate the flow rate, the following equation is adopted:

$$Q = const. \times \sqrt{\Delta p}, \tag{2}$$

where:

Δp is the differential pressure between the pressure of fluid inside the nozzle and ambient pressure; this differential pressure is directly obtained from the sensor.

Using table data for the flow at $\Delta p = 300\ kPa$, it is possible to express:

$$Q(\Delta p) = Q(300\ kPa) \times \sqrt{\frac{\Delta p}{300\ kPa}}. \tag{3}$$

For choosing the optimal set of spraying parameters, both the following conditions should be met:

- Average coverage $\bar{c}$ (%) should be acceptable (not too low, and with no or very few insufficient papers, Figure 4).
- Coverage in liters per hectare $C\left(\frac{L}{ha}\right)$ should be as low as possible.

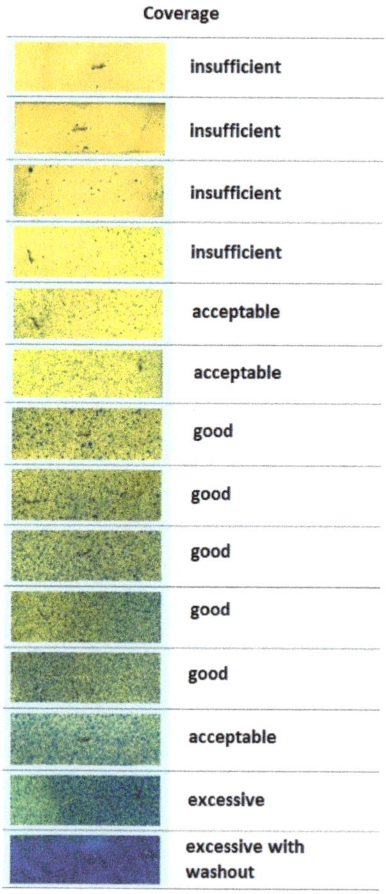

	Coverage
	insufficient
	insufficient
	insufficient
	insufficient
	acceptable
	acceptable
	good
	good
	good
	good
	good
	acceptable
	excessive
	excessive with washout

Figure 4. Degree of coverage of water sensitive papers [19].

It is intuitively clear that these parameters are, roughly speaking, inversely correlated.

3. Results and Discussion

The results of all the 37 tests are shown in Tables A1 and A2 (Appendix A) and analyzed according to the different parameters defined in the previous section: $N_{sufficient}$, $N_{insufficient}$, $\overline{c}$, σ_c, $\overline{d}$, σ_d, C. Columns are defined by these parameters, and for each test three rows were reserved for upper leaves, lower leaves, and ground.

Results for the water-sensitive samples collected from the ground showed that the spraying was carried out in all conditions at an almost complete coverage, thus suggesting not to focus on this information. This was expected, so the third ('ground') row of each test put in the table is not significant.

After applying the criterion for the number of "insufficient papers", only 10 tests remained for further consideration, which accounts for roughly 250 samples. The mixture (water) amount ranged from the very promising value of 265.5 L/ha to standard treatment of 957.6 L/ha (i.e., very close to the theoretical 1000 L/ha). Considering this work aimed to show the possibility to reduce significantly the pesticide distribution while maintaining good or even improving spraying results, it has been decided to go deep only in those tests involved in a distribution of less than 500 L/ha. The relevant results are shown in Table 1; to offer full insights, in Table 2 all the parameters affecting the results are shown.

Table 1. The selected results from the first (nos. 9, 13, 14) and the second test session (nos. 17, 21, 24, 27, 34).

Test	N_{suf}	N_{insuf}	$\overline{c}$ (%)	σ_c (%)	$\overline{d}$	σ_d	C (l/ha)
9-upper	25	0	85.8	12.0	4.67	4.15	426.0
9-lower	24	1	78.5	18.3	6.68	6.20	426.0
9-ground	24	0	96.6	1.5	0.86	1.49	426.0
13-upper	24	0	84.0	22.7	6.15	11.55	438.0
13-lower	21	3	81.6	15.0	6.25	5.04	438.0
13-ground	23	0	95.4	4.7	1.73	4.23	438.0
14-upper	23	0	95.2	4.4	1.39	1.82	469.6
14-lower	21	1	84.1	15.7	6.73	10.54	469.6
14-ground	21	0	97.0	0.5	0.42	0.27	469.6
17-upper	24	0	75.0	27.6	9.50	13.41	441.8
17-lower	22	1	81.5	26.4	7.00	14.33	441.8
17-ground	25	0	95.1	0.8	1.14	0.62	441.8
21-upper	24	0	67.1	28.3	11.41	9.55	435.3
21-lower	21	4	68.1	29.5	13.31	14.26	435.3
21-ground	24	0	94.1	2.9	1.94	2.32	435.3
24-upper	25	0	95.7	2.2	1.47	1.70	456.7
24-lower	21	0	95.8	2.8	1.52	1.59	456.7
24-ground	24	0	96.4	1.0	1.35	0.89	456.7
27-upper	22	2	62.4	32.7	16.94	17.40	265.5
27-lower	22	2	77.8	29.1	8.90	14.47	265.5
27-ground	20	0	94.7	4.4	2.61	3.22	265.5
34-upper	22	1	76.7	28.7	5.78	6.94	284.3
34-lower	21	2	79.7	26.9	7.37	12.09	284.3
34-ground	24	0	96.7	0.6	0.68	0.27	284.3

Table 2. Parameters from the tests 9, 13, 14, 17, 21, 24, 27 and 34.

Test Number	9	13	14	17	21	24	27	34
Height (cm)	53	27	27	37	49	49	49	65
Nozzle distance (cm)	40	25	25	25	40	50	50	50
Number of nozzles	3	3	3	3	3	2	2	2
Nozzle type (l/min) at 300 kPa	0.8	0.8	0.8	0.8	0.8	0.8	0.6	1
Nozzle inclination (°)	0	20	−20	0	−20	0	0	−20
Air inclination (°)	30	30	30	30	0	30	30	0
Speed (km/h)	2.25	4.15	3.86	4.32	2.51	3.12	3.48	4.00
Pressure (mbar)	1914	2694	2670	2964	2490	6614	4950	2694
Flow (l/min)	0.64	0.76	0.75	0.80	0.73	1.19	0.77	0.95
Average plant height (cm)	7	7	7	14	14	14	14	14
PWM [1] motor (%)	100	100	100	100	68	80	80	100
PWM [1] pump (%)	77	55	55	60	52	84	52	46

[1] pulse width modulation.

In Table 1, some runs show a slightly higher standard deviation σ_c. Although they have very good coverages $\bar{c}$, these values may suggest that by repeating the tests with the same parameters different results (possibly worse) could be obtained. To select the optimal parameters, we could have used the simple index of performance defined as $\eta = \frac{c}{C}$ could have been used. Although this makes sense as it represents the percentage coverage c (%) reached with a certain amount in L/ha, it is necessary to consider that the variance plays an important role. For this reason, we excluded to the time being very promising results (i.e., less than 300 L/ha) as they need further investigation.

Staying on the safe side, we can reliably state that with 420–470 L/ha it was possible to obtain excellent coverage of upper and lower leaves in both test sessions (for example, tests nos. 9 and 14 from the first session, with an average plant height of 7 cm and test no. 24 from the second session, with an average plant height of 14 cm).

Considering a theoretical mixture amount of 1000 L/ha, these numbers show that it is possible to optimize spraying parameters with a reduction of more than 50%. Going down to less than 470 L/ha, coverage results are however excellent, and still remain promising opportunities to reduce the usage below 300 L/ha, starting from the parameters of tests nos. 27 and 34 (Table 2).

4. Conclusions

Given that the commonly distributed mixture (amount of spray liquid for pesticide treatments) for the cultivation of ready-to-eat salads is 1000 L/ha, the tests carried out produced extremely promising results. Through the optimization of setting parameters, in the first life stage of the plant it is possible to decrease that amount up to 426 L/ha, maintaining an excellent coverage of the vegetation, with a very low wasting of PPPs to the ground. In a further stage of the plants growth, the optimal amount turned out to be 460 L/ha, still more than half of the standard.

The solution described will result in many benefits, ranging from environmental to economic sustainability. First, the overall treatment requires 55% less pesticide product and, consequently, chemicals will not be wasted anymore, avoiding soil and groundwater pollution. Furthermore, the usage of an electric rover replacing the diesel tractor will decrease the pollutant gas emission, also avoiding fossil fuel consumption and reducing CO_2 emissions.

Future studies should strive to explore additional improvements in pesticide usage with other parameters settings as, from the tests, very promising results have been obtained also below the proposed setting reaching 265.5 L/ha use.

Author Contributions: Conceptualization, D.F., G.F., M.M. and D.P.; Data curation, D.F., S.S., G.F., L.F., S.P., L.B.B., S.Š., F.S., G.C. and M.M.; Formal analysis, D.F., S.S., L.F., S.P., L.B.B., S.Š., F.S. and G.C.; Investigation, D.F., S.S., G.F., L.F., S.P., L.B.B. S.Š., F.S., G.C. and M.M.; Methodology, D.F., G.F., M.M. and D.P.; Project administration, D.F., G.F., M.M. and D.P.; Resources, D.F., L.E.G., G.F., M.M. and D.P.; Software, G.F., L.F., S.P., L.B.B., S.Š., F.S., G.C. and M.M.; Supervision, D.F., S.S., G.F., M.M. and D.P.; Validation, D.F., S.S., L.E.G., G.F., M.M. and D.P.; Visualization, D.F., S.S., L.E.G., G.F., M.M. and D.P.; Writing—original draft, D.F., S.S., L.E.G., M.M. and D.P.; Writing—review & editing, D.F., S.S., L.E.G., G.F., M.M. and D.P. All authors have read and agreed to the published version of the manuscript.

Funding: This research received no external funding.

Acknowledgments: We are grateful to "Azienda Agricola C.na Baciocca of the Università degli Studi di Milano" and to "AIR Lab—the Artificial Intelligence and Robotics Lab of Politecnico di Milano" for the facilities that have been made available to us to develop this project.

Conflicts of Interest: The authors declare no conflict of interest.

Appendix A

Table A1. Results of all the tests.

Test	N_{suf}	N_{insuf}	$\bar{c}$ (%)	σ_c (%)	$\bar{d}$	σ_d	C (*l/ha*)
1-upper	25	0	94.2	3.2	1.89	1.58	952.0
1-lower	8	12	31	19.9	22.62	24.80	952.0
1-ground	24	0	96.8	1.1	0.65	0.88	952.0
2-upper	25	0	85.8	12.0	4.67	4.15	426.0
2-lower	24	1	78.5	18.3	6.68	6.20	426.0
2-ground	24	0	96.6	1.5	0.86	1.49	426.0
3-upper	25	0	77.5	18.8	9.11	6.96	662.4
3-lower	11	13	36.3	21.0	43.09	21.63	662.4
3-ground	25	0	95.5	2.7	1.34	1.87	662.4
4-upper	23	2	38.9	17.8	12.78	9.03	435.0
4-lower	8	15	18.7	10.0	22.44	10.65	435.0
4-ground	25	0	96.7	0.8	0.66	0.47	435.0
5-upper	22	1	45.9	16.6	7.82	5.40	466.2
5-lower	8	15	35.4	23.3	13.72	9.02	466.2
5-ground	25	0	96.6	1.0	0.68	0.65	466.2
6-upper	25	0	78.6	14.3	4.54	3.27	447.9
6-lower	5	19	20.6	9.8	11.62	9.08	447.9
6-ground	25	0	96.6	1.4	0.87	0.99	447.9
7-upper	22	1	45.9	16.5	8.02	5.49	584.0
7-lower	8	15	35.2	23.3	13.79	9.24	584.0
7-ground	25	0	96.6	1.0	0.68	0.65	584.0
8-upper	25	0	78.8	19.5	6.32	6.35	622.5
8-lower	19	4	62.0	26.5	14.45	11.69	622.5
8-ground	25	0	96.7	1.5	0.67	0.69	622.5
9-upper	25	0	85.8	12.0	4.67	4.15	426.0
9-lower	24	1	78.5	18.3	6.68	6.20	426.0
9-ground	24	0	96.6	1.5	0.86	1.49	426.0
10-upper	25	0	75.9	28.3	5.46	6.21	492.2
10-lower	18	5	55.0	27.1	13.86	11.85	492.2
10-ground	24	0	96.9	1.0	0.47	0.47	492.2

Table A1. *Cont.*

Test	N_{suf}	N_{insuf}	$\bar{c}$ (%)	σ_c (%)	$\bar{d}$	σ_d	*C (l/ha)*
11-upper	22	3	62.0	17.8	9.97	6.44	306.8
11-lower	7	18	17.5	14.6	31.15	25.59	306.8
11-ground	24	0	97.3	0.6	0.37	0.24	306.8
12-upper	24	1	73.1	30.4	8.73	11.65	440.0
12-lower	7	18	53.8	25.8	15.37	11.27	440.0
12-ground	21	0	97.1	0.6	0.48	0.33	440.0
13-upper	24	0	84.0	22.7	6.15	11.55	438.0
13-lower	21	3	81.6	15.0	6.25	5.04	438.0
13-ground	23	0	95.4	4.7	1.73	4.23	438.0
14-upper	23	0	95.2	4.4	1.39	1.82	469.6
14-lower	21	1	84.1	15.7	6.73	10.54	469.6
14-ground	21	0	97.0	0.5	0.42	0.27	469.6
15-upper	21	1	57.8	30.8	13.39	13.67	438.0
15-lower	18	5	75.4	24.6	8.47	12.24	438.0
15-ground	25	0	91.6	16.8	4.17	14.36	438.0
16-upper	14	10	34.9	18.5	14.71	10.79	354.8
16-lower	7	12	21.6	12.2	29.80	17.04	354.8
16-ground	24	0	94.1	1.8	2.01	1.26	354.8
17-upper	24	0	75.0	27.6	9.50	13.41	441.8
17-lower	22	1	81.5	26.4	7.00	14.33	441.8
17-ground	25	0	95.1	0.8	1.14	0.62	441.8
18-upper	20	4	49.2	25.0	13.64	10.01	404.7
18-lower	15	5	36.7	23.1	24.63	18.50	404.7
18-ground	25	0	94.9	1.1	1.34	1.04	404.7
19-upper	18	6	49.9	31.7	13.50	13.98	432.5
19-lower	15	5	34.9	28.2	19.18	16.62	432.5
19-ground	25	0	94.2	1.9	1.70	1.53	432.5
20-upper	16	8	47.8	34.2	7.68	5.25	358.8
20-lower	10	15	28.5	26.8	24.38	17.02	358.8
20-ground	24	0	94.1	2.2	2.02	1.56	358.8
21-upper	24	0	67.1	28.3	11.41	9.55	435.3
21-lower	21	4	68.1	29.5	13.31	14.26	435.3
21-ground	24	0	94.1	2.9	1.94	2.32	435.3
22-upper	16	4	54.9	32.0	15.36	16.23	422.4
22-lower	20	3	56.8	33.2	13.84	14.64	422.4
22-ground	25	0	92.6	2.8	2.22	1.49	422.4
23-upper	23	0	94.0	4.3	2.03	2.37	379.8
23-lower	23	0	94.4	4.8	3.16	3.37	379.8
23-ground	25	0	93.1	4.6	3.68	3.77	379.8
24-upper	25	0	95.7	2.2	1.47	1.70	456.7
24-lower	21	0	95.8	2.8	1.52	1.59	456.7
24-ground	24	0	96.4	1.0	1.35	0.89	456.7
25-upper	23	0	93.3	13.6	2.06	4.21	347.1
25-lower	23	0	92.0	15.2	2.92	4.72	347.1
25-ground	23	0	96.0	1.4	1.65	1.13	347.1
26-upper	16	9	38.9	28.8	22.23	18.47	247.9
26-lower	19	5	42.2	30.3	21.65	13.09	247.9
26-ground	22	0	94.3	4.9	2.24	3.50	247.9

Table A1. *Cont.*

Test	N_{suf}	N_{insuf}	$\bar{c}$ (%)	σ_c (%)	$\bar{d}$	σ_d	C (l/ha)
27-upper	22	2	62.4	32.7	16.94	17.40	265.5
27-lower	22	2	77.8	29.1	8.90	14.47	265.5
27-ground	20	0	94.7	4.4	2.61	3.22	265.5
28-upper	23	2	72.1	30.4	9.67	14.47	380.5
28-lower	17	5	53.8	34.8	20.35	21.11	380.5
28-ground	20	0	96.6	0.5	0.80	0.38	380.5
29-upper	16	6	72.5	20.8	13.23	12.99	344.2
29-lower	22	2	82.0	23.4	8.69	14.13	344.2
29-ground	25	0	96.6	0.8	0.94	0.53	344.2
30-upper	25	0	83.3	22.0	5.44	8.62	957.6
30-lower	21	2	86.2	20.5	4.68	7.34	957.6
30-ground	25	0	96.9	0.7	0.55	0.33	957.6
31-upper	19	4	70.2	21.8	7.83	6.74	650.4
31-lower	15	9	46.0	28.4	19.45	17.35	650.4
31-ground	24	0	94.7	7.2	1.60	3.08	650.4
32-upper	22	3	79.6	26.6	8.01	13.98	697.3
32-lower	16	9	71.8	32.2	14.23	22.95	697.3
32-ground	22	0	94.2	5.3	1.22	0.78	697.3
33-upper	22	3	53.8	25.1	9.52	10.19	440.5
33-lower	14	10	42.4	29.5	10.24	9.05	440.5
33-ground	22	0	96.1	1.5	0.87	1.07	440.5
34-upper	22	1	76.7	28.7	5.78	6.94	284.3
34-lower	21	2	79.7	26.9	7.37	12.09	284.3
34-ground	24	0	96.7	0.6	0.68	0.27	284.3
35-upper	22	2	60.8	34.5	8.82	11.26	372.8
35-lower	14	11	48.2	30.9	23.40	16.06	372.8
35-ground	24	0	97.0	0.4	0.60	0.26	372.8
36-upper	21	3	54.4	28.3	9.08	13.00	192.0
36-lower	12	8	51.7	37.7	18.05	23.76	192.0
36-ground	24	0	95.6	4.8	1.48	3.86	192.0
37-upper	18	6	60.8	24.2	13.68	13.62	330.9
37-lower	13	10	61.3	30.7	12.27	12.38	330.9
37-ground	21	0	96.1	1.0	1.00	0.55	330.9

Table A2. Selected parameters of all 37 tests.

Test Number	**1**	**2**	**3**	**4**	**5**	**6**	**7**	**8**
Height (cm)	65	65	65	65	65	53	53	53
Nozzle distance (cm)	50	50	50	50	50	40	40	40
Number of nozzles	2	2	2	2	2	3	3	3
Nozzle type (l/min) at 300 kPa	2.4	1.6	1.6	1	0.8	0.8	0.8	0.8
Nozzle inclination (°)	2	2	−22	−22	−22	−22	−22	−22
Air inclination (°)	0	0	0	0	0	0	0	0
Speed (km/h)	2.45	2.45	2.50	2.51	2.35	2.45	2.45	2.00
Pressure (kPa)	197.5	216.7	223.2	248.7	390.0	251.8	428.1	322.9
Flow (l/min)	1.95	1.36	1.38	0.91	0.91	0.73	0.96	0.83
Average plant height (cm)	7	7	7	7	7	7	7	7
PWM [1] **motor** (%)	100	100	100	100	100	100	100	80
PWM [1] **pump** (%)	77	60	60	45	52	52	77	62

Test Number	**9**	**10**	**11**	**12**	**13**	**14**	**15**	**16**
Height (cm)	53	53	53	27	27	27	27	37
Nozzle distance (cm)	40	50	50	50	25	25	25	25

Table A2. *Cont.*

Test Number	1	2	3	4	5	6	7	8
Number of nozzles	3	3	2	2	3	3	3	3
Nozzle type (l/min) at 300 kPa	0.8	0.8	0.8	0.8	0.8	0.8	0.8	0.8
Nozzle inclination (°)	0	0	0	0	20	−20	−20	−20
Air inclination (°)	30	30	30	30	30	30	30	30
Speed (km/h)	2.25	2.25	3.60	4.15	4.15	3.86	3.91	3.86
Pressure (kPa)	191.4	399.3	397.0	271.8	269.4	267.0	239.0	152.4
Flow (l/min)	0.64	0.92	0.92	0.76	0.76	0.75	0.71	0.57
Average plant height (cm)	7	7	7	7	7	7	14	14
PWM [1] **motor** (%)	100	100	73	100	100	100	100	100
PWM [1] **pump** (%)	77	84	84	55	55	55	55	55

Test Number	17	18	19	20	21	22	23	24
Height (cm)	37	37	37	49	49	49	49	49
Nozzle distance (cm)	25	25	25	50	40	50	50	50
Number of nozzles	3	3	3	2	3	2	2	2
Nozzle type (l/min) at 300 kPa	0.8	0.8	0.8	0.8	0.8	1.0	1.2	0.8
Nozzle inclination (°)	0	45, −45	45, −45	−20	−20	−20	−20	0
Air inclination (°)	30	30	30	0	0	0	0	24
Speed (km/h)	4.32	3.86	3.62	2.51	2.51	2.70	3.97	3.12
Pressure (kPa)	296.4	198.3	199.9	264.3	249.0	271.0	329.0	661.4
Flow (l/min)	0.80	0.65	0.65	0.75	0.73	0.95	1.26	1.19
Average plant height (cm)	14	14	14	14	14	14	14	14
PWM [1] **motor** (%)	100	100	100	68	68	68	100	80
PWM [1] **pump** (%)	60	45	45	39	52	47	60	84

Test Number	25	26	27	28	29	30	31	32
Height (cm)	49	49	49	49	65	65	65	65
Nozzle distance (cm)	50	50	50	25	50	50	50	50
Number of nozzles	2	2	2	3	2	2	2	2
Nozzle type (l/min) at 300 kPa	0.8	0.6	0.6	0.6	0.6	2.4	1.6	1.6
Nozzle inclination (°)	0	0	0	45, −45	20	0	0	−20
Air inclination (°)	30	30	30	30	30	0	0	0
Speed (km/h)	3.18	3.38	3.48	4.15	3.00	2.57	2.45	2.35
Pressure (kPa)	395.6	405.0	495.0	361.5	617.1	219.3	207.4	218.1
Flow (l/min)	0.92	0.70	0.77	0.66	0.86	2.05	1.33	1.36
Average plant height (cm)	14	14	14	14	14	14	14	14
PWM [1] **motor** (%)	80	90	80	100	100	62	62	62
PWM [1] **pump** (%)	52	45	52	56	65	75	56	58

Test Number	33	34	35	36	37
Height (cm)	65	65	49	49	49
Nozzle distance (cm)	50	50	50	50	50
Number of nozzles	2	2	2	2	2
Nozzle type (l/min) at 300 kPa	0.8	1.0	1.2	0.6	0.8
Nozzle inclination (°)	−20	−20	−20	−20	0
Air inclination (°)	0	0	0	30	30
Speed (km/h)	2.45	4.00	4.06	3.72	3.29
Pressure (mbar)	380.6	269.4	331.4	2959	3865
Flow (l/min)	0.90	0.95	1.26	0.60	0.91
Average plant height (cm)	14	14	14	14	14
PWM [1] **motor** (%)	62	100	100	100	80
PWM [1] **pump** (%)	51	46	60	36	52

[1] pulse width modulation.

References

1. How to Feed the World 2050, High-Level Expert Forum FAO. Available online: http://www.fao.org/fileadmin/templates/wsfs/docs/Issues_papers/HLEF2050_Global_Agriculture.pdf (accessed on 11 December 2020).
2. The International Code of Conduct on Pesticides Management, FAO and WHO. Available online: http://www.fao.org/3/a-i3604e.pdf (accessed on 11 December 2020).
3. Pesticides, European Commission. Available online: https://ec.europa.eu/food/plant/pesticides_en (accessed on 14 December 2020).
4. Statistics on Pesticides Use in Agriculture, 1990–2018, FAO Environmental Statistics. Available online: http://www.fao.org/economic/ess/environment/data/pesticides-use/en/ (accessed on 12 December 2020).
5. Directive 2009/128/EC of the European Parliament and the Council of 21 October 2009. Available online: https://eur-lex.europa.eu/legal-content/EN/TXT/HTML/?uri=CELEX:02009L0128-20091125&from=EN (accessed on 14 December 2020).
6. Pesticides Use, FAOSTAT. Available online: http://www.fao.org/faostat/en/#home (accessed on 12 December 2020).

7. Agri-Environmental Indicator—Consumption of Pesticides, Eurostat. Available online: https://ec.europa.eu/eurostat/statistics-explained/index.php?title=Agri-environmental_indicator_-_consumption_of_pesticides#Data_sources (accessed on 12 December 2020).
8. Distribuzione per uso Agricolo dei Prodotti Fitosanitari (Erbicidi, Fungicidi, Insetticidi, Acaricidi e Vari), ISTAT. Available online: https://annuario.isprambiente.it/ada/downreport/html/6940 (accessed on 13 December 2020).
9. Italian Agriculture in Figures 2017, CREA. Available online: https://www.crea.gov.it/documents/68457/0/Itaconta+2017_+ING_DEF_WEB2.pdf/1218e51d-0bf5-03d0-3089-0a56af2e26d5?t=1559117850834 (accessed on 14 December 2020).
10. Gullino, M.L.; Gilardi, G.; Garibaldi, A. Ready-to-eat salad crops: A plant pathogen's heaven. *Plant Dis.* **2019**, *103*, 2153–2170. [CrossRef] [PubMed]
11. Casati, D.; Baldi, L. L'importanza economica del comparto della IV gamma. In *Le Malattie degli Ortaggi di IV Gamma*; Sanninoand, L., Espinosa, B., Eds.; Inteli Press: Battipaglia, Italy, 2016; pp. 19–30.
12. Intel RealSense Depth Camera d435i. Available online: https://www.intelrealsense.com/depth-camera-d435i/ (accessed on 16 September 2020).
13. Intel RealSense SDK 2.0 Library. Available online: https://github.com/IntelRealSense/librealsense (accessed on 16 September 2020).
14. Intel RealSense SDK 2.0 Library: Post-Processing Filters. Available online: https://github.com/IntelRealSense/librealsense/blob/master/doc/post-processing-filters.md (accessed on 16 September 2020).
15. Open-Source Computer Vision Library. Available online: https://github.com/opencv/opencv (accessed on 21 September 2020).
16. OpenCV: Color Detection and Segmentation Example. Available online: https://www.learnopencv.com/invisibility-cloak-using-colordetection-and-segmentation-with-opencv/ (accessed on 17 September 2020).
17. RGB Color Model. Available online: https://en.wikipedia.org/wiki/RGB_color_model (accessed on 17 September 2020).
18. HSV Color Model. Available online: https://en.wikipedia.org/wiki/HSL_and_HSV (accessed on 21 September 2020).
19. Quaderno: Agricoltura Responsabile. Syngenta. 2014. Available online: https://www.syngenta.it/file/5031/download?token=NyHxitSm (accessed on 25 August 2020).

MDPI
St. Alban-Anlage 66
4052 Basel
Switzerland
Tel. +41 61 683 77 34
Fax +41 61 302 89 18
www.mdpi.com

Applied Sciences Editorial Office
E-mail: applsci@mdpi.com
www.mdpi.com/journal/applsci

www.ingramcontent.com/pod-product-compliance
Lightning Source LLC
LaVergne TN
LVHW070623170726
843515LV00004B/160

* 9 7 8 3 0 3 6 5 1 3 0 3 4 *